Fundamentals of Mechanical Vibrations

Fundamentals of Mechanical Vibrations

Matthew Hussey

*Physics Department,
Dublin Institute of Technology*

© Matthew Hussey 1983

All rights reserved. No part of this publication may be reproduced or transmitted, in any form or by any means, without permission.

First published 1983 by
THE MACMILLAN PRESS LTD
London and Basingstoke
Companies and representatives
throughout the world

Typeset in 10/12 Times by
Photo-Graphics, Honiton, Devon

Printed in Hong Kong

ISBN 0 333 32436 6
ISBN 0 333 32437 4 pbk

The paperback edition of the book is sold subject to the condition that it shall not, by way of trade or otherwise, be lent, resold, hired out, or otherwise circulated without the publisher's prior consent in any form of binding or cover other than that in which it is published and without a similar condition including this condition being imposed on the subsequent purchaser.

For my Mother and Father

Contents

Preface   xi

List of Symbols   xii

1 Fundamental Mechanical Quantities   1
    1.1 Introduction and Objectives   1
    1.2 Force and Torque   1
    1.3 Hooke's Law   8
    1.4 Work and Mechanical Energy   13
    1.5 Units   17
    1.6 Experiments   17
    Problems   20
    Further Reading   22

2 Free Harmonic Oscillator   23
    2.1 Introduction and Objectives   23
    2.2 The Simple Harmonic Oscillator   23
    2.3 Graphical Display of Simple Harmonic Motion   27
    2.4 Forces in the Oscillator   28
    2.5 Energy in the System   28
    2.6 Initial Conditions   30
    2.7 Other Simple Harmonic Oscillators   32
    2.8 Electrical analogues of the Simple Harmonic Oscillator   37
    2.9 Experiments   40
    Problems   41
    Further Reading   43

3 Damped Free Harmonic Oscillator   44
    3.1 Introduction and Objectives   44
    3.2 Friction and Damping   44
    3.3 Damped Harmonic Oscillator   51
    3.4 Three Different Types of Behaviour   52
    3.5 Energy in the Damped System   58
    3.6 Functional Parameters of the Underdamped System   62
    3.7 Electrical Analogue Circuits   64

	3.8	Experiments	66
		Problems	68
		Further Reading	70

4 Sinusoidal Driving of the Damped Harmonic Oscillator 72
- 4.1 Introduction and Objectives — 72
- 4.2 Sinusoidally Forced Harmonic Oscillator — 72
- 4.3 Mechanical Impedance and Compliance — 79
- 4.4 Energy and Power in the Driven Oscillator — 83
- 4.5 Power Resonance — 87
- 4.6 Electrical Analogue Circuits — 92
- 4.7 Experiments — 93
- Problems — 94
- Further Reading — 96

5 Harmonic Analysis and Transform Techniques in Damped Harmonic Oscillator Systems — 97
- 5.1 Introduction and Objectives — 97
- 5.2 Fourier Theory of Harmonic Analysis — 97
- 5.3 Truncated Fourier Series — 104
- 5.4 Pulse Trains, Isolated Pulses and Shocks — 107
- 5.5 Laplace Transforms — 111
- 5.6 Linearity and Superposition — 116
- 5.7 Application of Laplace Transform Technique to a Linear System — 117
- 5.8 Impulse Response and System Transfer Function — 119
- 5.9 Experiments — 124
- Problems — 124
- Further Reading — 126

6 Coupled Harmonic Oscillators — 127
- 6.1 Introduction and Objectives — 127
- 6.2 Coupled Oscillating Systems — Modes — 127
- 6.3 Energy in the Coupled System — 134
- 6.4 Resonances in the Coupled System — 136
- 6.5 Electrical Analogues — 139
- 6.6 Experiments — 141
- Problems — 142
- Further Reading — 144

7 Non-discrete-element Mechanical Vibratory Systems — 145
- 7.1 Introduction and Objectives — 145
- 7.2 Simple Harmonic Oscillator with Massy Spring — 145
- 7.3 Compound Pendulum — 147

CONTENTS ix

	7.4	Vibrating String	149
	7.5	Solid Rod in Transverse or Flexural Vibration	154
	7.6	Longitudinal Vibrations in a Solid Rod	159
	7.7	Vibrating Circular Membrane	163
	7.8	Vibrations in Air Columns	168
	7.9	Electrical Analogue of Distributed Oscillators	171
	7.10	Experiments	173
		Problems	176
		Further Reading	179
8	**The Piezoelectric Effect**	**180**
	8.1	Introduction and Objectives	180
	8.2	Basic Aspects of the Piezoelectric Effect	180
	8.3	The Piezoelectric Relationships	182
	8.4	Alternative Forms of the Piezoelectric Relationships	185
	8.5	Wave Equation in a Piezoelectric Slab	190
	8.6	Experiments	197
		Problems	197
		Further Reading	198
9	**Non-linear Vibratory Systems**	**199**
	9.1	Introduction and Objectives	199
	9.2	Non-linear Spring	199
	9.3	Non-linear Spring and Mass under Harmonic External Force	204
	9.4	Non-linear Damping in a Freely Vibrating Simple Oscillator	207
	9.5	Oscillator with Hysteretic Damping under Harmonic Force	211
	9.6	Static Friction Damping in an Oscillator	214
	9.7	Experiments	217
		Problems	219
		Further Reading	220
10	**Vibration Measurement and Analysis**	**222**
	10.1	Introduction and Objectives	222
	10.2	Effects of Vibrations in Mechanical Systems	222
	10.3	Measurement Equipment	224
	10.4	Vibration Transducers, Accelerometer and Vibrometer	225
	10.5	Other Displacement, Strain and Force Transducers	231
	10.6	Analysis of Periodic Vibrations	233
	10.7	Random Vibrations	235
	10.8	Shocks/Pulsed Vibrations	244

	10.9	Experiments	244
		Problems	245
		Further Reading	247

11 Vibration Isolation and Control — 248
- 11.1 Introduction and Objectives — 248
- 11.2 Sources of Vibrations — 248
- 11.3 Vibration Prevention — 250
- 11.4 Isolation of Vibration Source — 252
- 11.5 Protection of Equipment against Vibrations — 257
- 11.6 Dynamic Vibration Absorbers — 258
- 11.7 Experiments — 263
- Problems — 264
- Further Reading — 266

12 Vibrations and the Human Body — 267
- 12.1 Introduction and Objectives — 267
- 12.2 Effects of Mechanical Vibrations on Human Beings — 267
- 12.3 The Body as a Mechanical System — 272
- 12.4 Ballistocardiography — 274
- 12.5 Pulsatile Wave Propagation along Arteries in the Body — 276
- 12.6 Experiments — 281
- Problems — 282
- Further Reading — 284

Answers to Problems — 283

Index — 293

Preface

A knowledge and appreciation of mechanical vibrations are fundamental to wide areas of physics, engineering and architecture. Naturally occurring vibrations are of interest to geophysicists, to civil engineers and to medical physicists. Guarding against the effects of excessive vibrations is a prime concern of design engineers, architects, civil engineers, structural and mechanical engineers, instrumentation physicists and biomedical engineers. The design and construction of vibrating machines is often the aim of mechanical engineers and acoustic engineers. The measurement and interpretation of vibrations are important tasks for maintenance engineers and physicists.

The aim of this book is to provide the basic groundwork in mechanical vibrations for these various disciplines. Practice in problem-solving is probably the single most powerful mechanism for learning and therefore it is strongly recommended that the student attempt all of the problems appended to each chapter. The value of the text will be greatly enhanced if its study is accompanied by a course of related laboratory experiments. Each chapter has a list of suggested laboratory exercises relating to the material in that chapter. At the end of each chapter there is a brief list of books for further reference. Whenever possible, the student should consult these in order to supplement and expand on the coverage in this text, to obtain alternative insights, to delve deeper into specific topics and to find additional problems.

Students embarking on this book should have the following prerequisites: differential and integral calculus, including partial differential equations; physics to approximately first-year college level; A.C. electricity, which could be studied concurrently; mechanical properties of materials, which could also be studied concurrently.

Thanks are due to Ms Elema Flanagan for doing much of the art work and to Ms Gerardine Keating for typing the first draft.

MATTHEW HUSSEY

List of Symbols

A	area	m²
	amplitude constant	—
a	amplitude constant	
	radius	m
α	coefficient of decay or decay constant	1/s
	attenuation constant	1/m
B	amplitude constant	—
	bulk modulus of elasticity	Pa
b	damping coefficient	N s/m or kg/s
	amplitude constant	—
β	propagation or wavelength constant	1/m
C	spring constant	N/m
	amplitude constant	—
	electrical capacitance	F
	inverse of compliance per metre of tube wall	Pa/m²
c	amplitude constant	—
	spring constant	N/m
	speed of propagation	m/s
	Young's modulus of elasticity	N/m² or Pa
D	dielectric displacement	C/m²
d	distance	m
	amplitude constant	—
	piezoelectric constant	C/N or m/V
δ	central deflection of beam	m
	logarithmic decrement	—
	Dirac delta function	—
	non-linear constant for spring	—
Δ	ΔL, increment of L	
E	Young's modulus of elasticity	N/m² or Pa
	energy	J
	electric field strength or potential gradient	V/m
e	piezoelectric constant	C/m² or N/V m
	eccentric radius	m
ϵ	dielectric constant	C/m V
F	force	N
	Fourier transform	—

LIST OF SYMBOLS

f	frequency	Hz
	instantaneous force	N
	general function	—
φ	angular displacement	rad
	phase angle	rad
Φ	maximum angular deflection	rad
G	force divided by mass	m/s^2
	Gibbs' free energy	J
	amplitude constant	—
g	acceleration due to gravity	= 9.81 m/s^2
	instantaneous force divided by mass	m/s^2
	piezoelectric constant	m^2/C or V m/N
	mass ratio in dynamic vibration absorber	—
γ	end deflection of cantilever	m
	phase angle	rad
	ratio of specific heats of gas	—
	complex propagation constant	1/m
H	heat	J
	system transfer function	—
	enthalpy	J
	amplitude constant	Pa
$\dot{H}$	rate of heat production	J/s
h	height	m
	impulse response	—
	piezoelectric constant	N/C or V/m
	hysteretic damping coefficient	N/m
η	coefficient of viscosity	kg/s m^2
I	mass moment of inertia	kg m^2
	area moment of inertia	m^4
	electrical current	A
i	instantaneous electrical current	A
$\dot{i}$	rate of change of electrical current	A/s
J	Bessel function	—
	electrical current density	A/m^2
j	$\sqrt{-1}$	—
K	constant	J
	integer	= 1, 2, 3...
k	constant	—
	integer	= 0, 1, 2...
	coupling coefficient of piezoelectric material	—
$\varkappa$	mass per unit length of string	kg/m
	compliance or compressibility	m^2/N
	non-linearity constant of spring	—

LIST OF SYMBOLS

L	angular momentum	kg m² rad/s
	length	m
	electrical inductance	H
	Laplace transform	—
$\mathscr{L}$	Laplace transform operation	—
$\mathscr{L}^{-1}$	Inverse Laplace transform operation	—
l	length	m
	integer	= 0, 1, 2 ...
λ	wavelength	m
M	mass	kg
	bending moment or torque	N m
m	integer	= 0, 1, 2 ...
μ	coefficient of static friction	—
	mass per unit area of membrane	kg/m²
Mom	linear momentum	kg m/s
N	Bessel function	—
n	shear modulus of elasticity	N/m² or Pa
	integer	= 0, 1, 2 ...
P	power	W
	pressure	Pa
	amplitude of mode displacement	m
	distribution function	—
	protection factor	—
p	constant	m
	instantaneous power	W
	amplitude constant	—
	mode displacement	m
	instantaneous pressure	Pa
	initial displacement	m
	probability density	—
$\dot{p}$	rate of change of instantaneous pressure	Pa/s
π	constant	= 3.14159...
ψ	phase angle	rad
	mode co-ordinate	m kg^{1/2}
	angular frequency	rad/s
Ψ	amplitude of mode co-ordinate	m kg^{1/2}
Q	quality factor or value	—
	2 × spectral density function	—
$\dot{Q}$	flow rate	m³/s
q	constant	m
	instantaneous electrical charge	C
	damping ratio	—
	amplitude constant	—
	initial velocity	m/s
$\dot{q}$	instantaneous flow rate	m³/s

LIST OF SYMBOLS

R	electrical resistance	Ω
	radial distance	m
	universal gas constant	= 8.31 J/mole K
r	radius	m
	radial axis	—
	ratio of frequency to natural frequency	—
ρ	density	kg/m³
	resistivity	Ω m
S	strain	—
	spectral density function	—
s	circumferential displacement	m
	transform parameter	—
	ratio of third to first harmonic amplitudes	—
$\dot{s}$	circumferential velocity	m/s
$\ddot{s}$	circumferential acceleration	m/s²
σ	force per unit length of periphery of membrane	N/m
	entropy	J/K
	standard deviation	—
T	stress	N/m² or Pa
	period of oscillation or of periodic function	s
	transmissibility	—
t	time	s
	temperature	°C
τ	time constant or relaxation time	s
	duration of pulse	s
	time variable	s
ϑ	angular displacement	rad
	angle	rad
	angular axis in cylindrical co-ordinates	—
	absolute temperature	K
$\dot{\vartheta}$	angular velocity	rad/s
$\ddot{\vartheta}$	angular acceleration	rad/s²
Tr	torque or turning moment	N m
U	step function	—
	internal energy	J
V	volume	m³
	electrical potential difference	V
v	instantaneous potential difference	V
w	distance	m
ω	angular velocity or angular frequency	rad/s
Wk	work	N m or J
X	Laplace transform of x	—
x	axis	—
	displacement along x-axis	m
	change in length of spring	m

LIST OF SYMBOLS

$\dot{x}$	velocity in x-direction	m/s
$\ddot{x}$	acceleration in x-direction	m/s²
ξ	particle displacement along x-axis	m
$\dot{\xi}$	particle velocity along x-axis	m/s
$\ddot{\xi}$	particle acceleration in x-direction	m/s²
Y	mechanical compliance	s/kg
y	axis	—
	displacement along y-axis	m
	distance	m
$\dot{y}$	velocity along y-direction	m/s
$\ddot{y}$	acceleration along y-direction	m/s²
Z	mechanical impedance	kg/s or N s/m
	electrical impedance	Ω
z	axis	—
	displacement along z-axis	m
$\dot{z}$	velocity along z-axis	m/s
$\ddot{z}$	acceleration along z-direction	m/s²
ζ	particle displacement along z-axis	m
$\dot{\zeta}$	particle velocity in z-direction	m/s
$\ddot{\zeta}$	particle acceleration in z-direction	m/s²

1 Fundamental Mechanical Quantities

1.1 Introduction and Objectives

Mechanical vibrations are the oscillatory motions, either continuous or transient, of objects and structures. In some instances they are purposeful and integral to the design of a machine as in a pneumatic drill or a reciprocating engine. In most instances, however, they are incidental or accidental and may impair the normal functioning of a structure or instrument.

Such vibrations enter into all aspects of the mechanical world and are therefore of interest to some extent in all fields of engineering science and physics. A knowledge of the fundamentals of mechanical vibrations is indispensable to practitioners of these varied technologies. The over-all aim of this book is to provide an introduction to the concepts and practical manifestations of mechanical vibrations geared to undergraduate and advanced technician diploma levels.

This first chapter recapitulates the basic ideas of motion, mechanical force and energy which underpin vibrations.

After reading this chapter the student should be able to

(a) define force and torque;
(b) derive the relationships between acceleration, velocity and displacement for linear motion and for angular motion;
(c) combine forces and torques vectorially;
(d) define Hooke's law, spring constant and Young's modulus;
(e) describe the spring characteristics for various types of spring;
(f) define mechanical work, energy and power;
(g) calculate kinetic energy in linearly and angularly moving bodies;
(h) calculate potential energy in various types of spring;
(i) calculate potential energy by virtue of position in the earth's gravitational field.

1.2 Force and Torque

Consider the case of linear, translational motion of an object along the

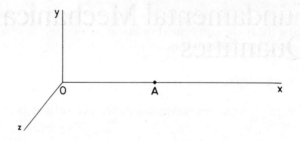

Figure 1.1 The three cartesian co-ordinates x, y and z, showing a displacement $x = OA$ along the x-axis.

x-axis of figure 1.1. Movement from the origin O, a distance x m, to A allows the displacement of the object to be written

$$\text{displacement} = + x \text{ m} . \tag{1.1}$$

Displacement is measured in metres (m). Retracing the path back to the origin needs a displacement of $- x$.

The rate of change of displacement with time is called the velocity of the body, thus

$$\text{velocity} = \dot{x} . \tag{1.2}$$

The unit of velocity is metres per second (m/s). Velocity from left to right is taken as positive while from right to left is negative.

The rate of change of velocity with time is the acceleration of the object, thus

$$\text{acceleration} = \ddot{x} , \tag{1.3}$$

and the relevant unit is metres per second per second (m/s²). Rate of increase of velocity is positive acceleration while rate of decrease of velocity is negative acceleration, deceleration or retardation.

Newton's second law states that when a force F, acts on an object of mass M kg, it causes a translational acceleration $\ddot{x}$, given by the following equation

$$F = M\ddot{x} . \tag{1.4}$$

If an object of mass M is being accelerated, then equation 1.4 asserts that it is being acted on by a force F. If M is in units of kilograms (kg) and $\ddot{x}$ in m/s², then F is in units of newtons (N).

FUNDAMENTAL MECHANICAL QUANTITIES

A more general definition of force is that it is the rate of change of the momentum of the body with time. For linear motion the momentum is the product of mass and velocity thus

$$\text{Mom} = M\dot{x} \ . \tag{1.5}$$

The units of momentum are kilogram metre per second (kg m/s). Thus force is given by

$$F = \frac{d}{dt}(M\dot{x})$$
$$= \dot{x}M' + M\ddot{x} \ . \tag{1.6}$$

When, as in most common situations, the mass is fixed equation 1.6 reduces to equation 1.4.

Displacement, velocity, momentum, acceleration and force are all vector quantities — each is characterised by both a magnitude and a direction. Diagrammatically, vector quantities are represented by an arrow, the length of the arrow being proportional to the vector magnitude and the direction of the arrow representing the vector orientation.

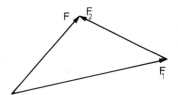

Figure 1.2 The vector summation of two forces F_1 and F_2 which act simultaneously on the same point of application is the force F.

An object experiencing two separate vectors simultaneously, such as two forces, F_1 and F_2 as in figure 1.2, behaves as if acted on by one force F, which is the vector sum of F_1 and F_2.

$$F = F_1 + F_2 \ . \tag{1.7}$$

The + sign here denotes vector summation and the operation is illustrated in figure 1.2.

Just as two or more force vectors can be combined into a single equivalent force vector, so also a single force vector may be decomposed into two or more component forces. Thus, as shown in figure 1.3, the force

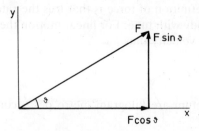

Figure 1.3 The force F which acts at an angle ϑ to the x-axis is equivalent to the vector sum of the x-axis projection of F, $F \cos \vartheta$, and the y-axis projection of F, $F \sin \vartheta$. Thus $F \cos \vartheta$ is the x-component of F while $F \sin \vartheta$ is the y-component of F.

F is equivalent to the vector sum of its x-component, $F \cos \vartheta$, and its y-component, $F \sin \vartheta$.

An important special case arises when the vector sum of the two forces acting through the centre of gravity of the body equals zero. Then from equation 1.7

$$F_1 = - F_2 . \tag{1.8}$$

Two such cases are possible and are illustrated in figure 1.4. The two equal but opposite forces are both directed either towards the centre of gravity, C, or away from C. This is the condition of equilibrium of forces. There being no net force, the object is not accelerated.

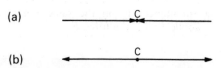

Figure 1.4 (a) Two equal and opposite forces subjecting the body with centre of gravity at C to uniaxial compression. There is no net force and so the body is in equilibrium. (b) Two equal and opposite forces which subject the body to tension. The body is in equilibrium.

Two equal and oppositely directed forces that do not act along the same line through an object, as in figure 1.5, produce rotational or angular motion of the mass. In rotation, each element of the body moves along the circumference of a circle centred on the axis of rotation. Thus in the

FUNDAMENTAL MECHANICAL QUANTITIES

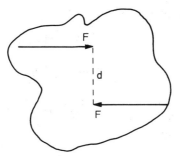

Figure 1.5 A body subjected to two equal, opposite but not co-linear forces. The net turning moment or torque Fd produces rotation of the body.

rectangular object of figure 1.6, which is rotating counter-clockwise about an axis through the point O, the element of the body at A, r m from O, travels along the circumference of the circle of radius r. When A undergoes the translation s, from A_0 to A_1, the radial line OA undergoes the angular displacement ϑ such that

$$s = r\vartheta \ . \tag{1.9}$$

Angular displacement is measured in radians (rad). The rate of change of angular displacement with time is the angular velocity.

$$\begin{aligned}\text{angular velocity} &= \dot{\vartheta} \\ &= \frac{1}{r}\dot{s} \ ,\end{aligned} \tag{1.10}$$

and is measured in radians per second (rad/s). The rate of change of angular velocity with time is the angular acceleration.

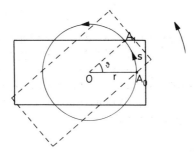

Figure 1.6 A rectangular body rotating counter-clockwise about an axis through O. Each point in the body, such as A, describes the circumference of a circle centred on O.

$$\text{angular acceleration} = \ddot{\vartheta}$$
$$= \frac{1}{r}\ddot{s}. \qquad (1.11)$$

The units of angular acceleration are radians per second per second (rad/s^2).

When the body is rotating at a given angular velocity $\dot{\vartheta}$, each elemental mass dM of the body has a translational velocity $\dot{s}$ and hence an elemental momentum d(Mom), thus

$$d(\text{Mom}) = dM\,\dot{s}$$
$$= \dot{\vartheta}\,r\,dM. \qquad (1.12)$$

The angular momentum of this elemental mass about O is dL given by

$$dL = r\,d(\text{Mom})$$
$$= \dot{\vartheta}\,r^2\,dM. \qquad (1.13)$$

The total angular momentum of the rotating body is L given thus

$$L = \dot{\vartheta}\int r^2\,dM$$
$$= \dot{\vartheta} I, \qquad (1.14)$$

where the integration is over the whole body and

$$I = \int r^2\,dM \qquad (1.15)$$

is the mass moment of inertia of the body about the axis through O.

In angular motion, the analogue of force is turning moment or torque Tr, which, with reference to figure 1.5, is defined as

$$\text{Tr} = Fd \qquad (1.16)$$

where F is the magnitude of the two equal and opposite forces and d is the perpendicular distance between their lines of application. Torque is thus measured in newton metres (N m). Such a torque is a vector quantity with a magnitude and a sense which may be clockwise (taken as negative) or counter-clockwise (positive). When a torque acts on an object, Newton's second law may be adapted thus

$$\text{Tr} = I\ddot{\vartheta} \qquad (1.17)$$

where the moment of inertia I is the analogue of mass in equation 1.4 and the angular acceleration $\ddot{\vartheta}$ is the analogue of translational acceleration in

FUNDAMENTAL MECHANICAL QUANTITIES

that equation. Two or more torque vectors acting simultaneously may be added vectorially. Two equal and opposite torques acting on the same body produce no net or unbalanced torque and hence no angular acceleration.

Example 1.1

A mass of 100 kg, initially at rest, experiences a steady force of 150 N for 5 s. What is the velocity of the body at the end of those 5 s and how far will it have travelled by then?

Solution

Here $x(0) = 0$ and $\dot{x}(0) = 0$. From equation 1.4

$$F = 150 = M\ddot{x} = 100\ddot{x} \ .$$

Hence $\ddot{x} = 1.5$, and so by integration

$$\dot{x} = 1.5t \ .$$

Thus when $t = 5$, $\dot{x} = 7.5$ m/s.
The displacement at that instant is given by another integration

$$x = \int \dot{x} \, dt = 0.5 \times 1.5 \, t^2 \, |_0^5$$
$$= 0.75 \times 25 = 18.75 \text{ m} \ .$$

Example 1.2

A flywheel of mass 1000 kg, all assumed to be evenly distributed along the rim 2 m from the axis of rotation, experiences a steady torque which after 1 minute brings it from rest to a rotational speed of 1000 rotations per minute. Find the torque.

Solution

Here $\vartheta(0) = 0$ and $\dot{\vartheta}(0) = 0$. Also the moment of inertia from equation 1.15 is

$$I = \int r^2 \, dM = r^2 M = 2^2 \times 10^3 = 4000 \text{ kg m}^2$$

After 60 s

$$\dot{\vartheta} = \frac{1000 \times 2\pi}{60} = 104.7 \text{ rad/s} = \ddot{\vartheta} t .$$

So then

$$\ddot{\vartheta} = \frac{104.7}{60} = 1.75 \text{ rad/s}^2 .$$

Therefore from equation 1.17 the torque is given by

$$\text{Tr} = I\ddot{\vartheta} = 7000 \text{ N m} .$$

1.3 Hooke's Law

Consider the cylinder of length L and cross-sectional area A to be subjected to two balanced forces as shown in figure 1.7. The pair of forces produces no acceleration of the body as a whole but stretches or elongates

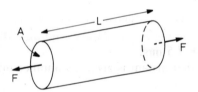

Figure 1.7 A uniform cylindrical specimen of length L m and cross-sectional area A m^2 is subjected to two equal and opposite tensile forces F.

it. These forces are tensile forces. The resultant increase in length of the cylinder is ΔL and for most materials of which the cylinder might be made there is a linear relationship between the magnitude of the counterbalanced forces F and ΔL. Thus

$$F = C (\Delta L) \tag{1.18}$$

where C is a constant. This equation is an expression of Hooke's law of elasticity. The constant C is called the elastic or spring constant of the cylinder subjected to axial forces.

If each of the two forces acts in the opposite direction $(-F)$ they tend to uniaxially compress the cylinder. They are then uniaxial compressive

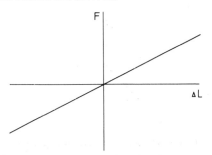

Figure 1.8 A linear relationship between force F and resultant change in length ΔL, for both tensile (positive) and uniaxial compressive (negative) forces, exemplifies Hooke's law.

forces and they tend to shorten the cylinder producing a negative ΔL. Hooke's law still pertains thus

$$-F = C(-\Delta L) . \tag{1.19}$$

Figure 1.8 summarises Hooke's law behaviour, that is, a linear relationship between F and the resulting ΔL of slope C and this characteristic straight line passes through the origin.

One further aspect of elasticity, which can not be depicted on figure 1.8, is reversibility, the property whereby, on removing the force completely, the elongation or shortening reduces to zero.

The spring constant C, in equations 1.18 and 1.19, depends on the cylinder geometry and on more basic properties of the material constituting the cylinder. To uncover the ingredients in C, consider again the loaded specimen of figure 1.7. Each successive cross-section of the cylinder along its length must support the full tensile force F. The distribution of this force over the cross-sectional area is called the stress T, where

$$T = \frac{F}{A} \tag{1.20}$$

and is measured in newtons per square metre (N/m^2) or pascals (Pa).

The resultant change in length per unit original length is the strain S, where

$$S = \frac{\Delta L}{L} \tag{1.21}$$

and has no dimensions.

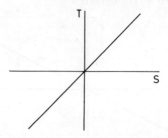

Figure 1.9 For Hookean, elastic behaviour the stress T in a material, in tension or uniaxial compression, is linearly related to resultant strain S.

For most materials the relationship between stress and resultant strain is linear as shown in figure 1.9 and as summarised in the following equation

$$T = ES \ . \tag{1.22}$$

Here E is a constant, Young's modulus of elasticity, which is a fundamental characteristic of the material of the cylinder. The units of E are N/m² or Pa. Clearly this equation is another version of Hooke's law of elasticity.

From equations 1.18 to 1.22 one can write for the cylinder in uniaxial loading

$$C = \frac{EA}{L} \ . \tag{1.23}$$

Thus the spring constant C is directly proportional to the cross-sectional area A, inversely proportional to the length of the cylinder L and directly proportional to the Young's modulus of the material of the cylinder E. When the value of C is large, the spring is said to be stiff or rigid. For a given geometry, the stiffness is set by the choice of material. In general the material and the cylinder dimensions may be chosen to set the value of C. A spring with a low value of C is said to be flexible.

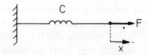

Figure 1.10 The conventional representation for a spring in which the change in length due to the application of a force F is the displacement x of the moving end of the spring.

FUNDAMENTAL MECHANICAL QUANTITIES

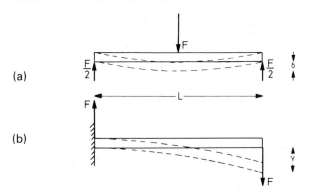

Figure 1.11 Two examples of beams functioning as springs. (a) A three-point-loaded beam in which the force F acts at the mid-point and the deflection of that mid-point is δ. (b) A cantilever beam with the force F acting at the outer extremity resulting in a deflection there of γ.

The conventional schematic representation of a spring of constant C being deformed along the x-axis is shown in figure 1.10. The origin of the x-axis is set at the end of the unstretched spring so that x then becomes the change in length. Hooke's law then becomes

$$F = Cx . \tag{1.24}$$

A number of different forms of spring are used in practical structures. The two beams shown in figure 1.11, (a) the three-point loaded beam and (b) the cantilever beam, are cases in point. Thus, for the former type of beam, if δ be the deflection of the centre-point of the beam, Hooke's law takes the form

$$F = \left(\frac{48\,EI}{L^3}\right)\delta \tag{1.25}$$

where E is Young's modulus for the material of the beam, L is the length of the beam and I is the area moment of inertia of the cross-section of the beam about the neutral axis of the beam (see section 7.5).

For the cantilever beam, loaded as shown in figure 1.11(b), Hooke's law may be written

$$F = \left(\frac{3\,EI}{L^3}\right)\gamma \tag{1.26}$$

where γ is the deflection of the end of the beam.

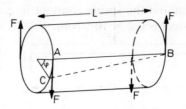

Figure 1.12 A cylindrical specimen of length L subjected to elastic torsion about the long axis. The two balanced torques acting at each end of the rod produce an angular twist of φ.

An alternative version of Hooke's law applies to the cylindrical specimen of material experiencing two equal but opposite torques about the long axis as shown in figure 1.12. The pair of torques cause a twist or torsion of the rod about the long axis. The point A is shifted to the location C, if the right-hand end is held rigidly in place. The angular deflection of the left-hand end is φ. Hooke's law in this case takes the form

$$\mathrm{Tr} = \left(\frac{nI}{L}\right)\varphi \qquad (1.27)$$

where L is the length of the rod, I is the area moment of inertia of the cross-section of the rod about its central or long axis and n is a constant, the shear modulus of elasticity, a characteristic of the material of the rod. Here torque is the analogue of force, angular deflection φ is the analogue of change in length and the spring constant analogue is (nI/L). Note that here too the spring constant incorporates geometry as well as a basic material parameter. The very commonly used helical spring is a special form of this torsion spring.

It should be borne in mind that Hooke's law applies to a good approximation to these various spring types only for relatively small deformations. As the force is increased, the force/deformation or stress/strain graph becomes non-linear. If a material is strained to this non-linear part of the stress/strain graph some permanent or irreversible strain can happen. Such non-elastic or plastic deformation is clearly a gross departure from Hookean elastic behaviour. Eventually if the force is increased to a sufficiently high level, the spring fractures.

Example 1.3

A rod of steel 1 m long and of square cross-section 1 cm × 1 cm is subjected to a force of 1000 N, (a) in tension and (b) in three-point-loading.

FUNDAMENTAL MECHANICAL QUANTITIES

Calculate the deflection and the spring constant in each case, given that Young's modulus for steel is 0.2×10^{12} N/m².

Solution

(a) Tension

$$A = 10^{-4} \text{ m}^2 .$$
$$\text{stress } T = 10^7 \text{ N/m}^2 .$$

Therefore the strain, from equation 1.22

$$S = \frac{T}{E} = \frac{10^7}{0.2 \times 10^{12}} = 50 \times 10^{-6} .$$

The elongation $\Delta L = LS = 50 \times 10^{-6}$ m .

The spring constant $C = \frac{F}{\Delta L} = \frac{10^3}{50 \times 10^{-6}} = 20 \times 10^6$ N/m .

(b) Three-point bending
Moment of inertia of the cross-section

$$I = \frac{(10^{-2})^4}{24} = \frac{10^{-8}}{24} \text{ m}^4 .$$

From equation 1.25, the central deflection δ is

$$\delta = \frac{FL^3}{3EI} = \frac{10^3 \times 1^3 \times 24}{48 \times 0.2 \times 10^{12} \times 10^{-8}}$$
$$= 0.25 \text{ m} .$$

Hence the spring constant

$$C = \frac{F}{\delta} = \frac{10^3}{0.25} = 4 \times 10^3 \text{ N/m} .$$

1.4 Work and Mechanical Energy

When the point of application of a net force F is moved a distance d then the force is said to do an amount of work Wk given by

$$\text{Wk} = Fd . \tag{1.28}$$

If F is in newtons and d is in metres, the unit of Wk is the joule (J). This amount of work done on the object increases the mechanical energy of the object by Wk J. If this object is a mass M, it is accelerated by the force. Consider the case of acceleration from rest at $x = 0$. Thus the acceleration is found from equation 1.4

$$\ddot{x} = \frac{F}{M} . \qquad (1.29)$$

Integrating each side of this equation gives

$$\dot{x} = \left(\frac{F}{M}\right) t + \dot{x}(0) . \qquad (1.30)$$

But $\dot{x}(0)$, the value of the velocity at the start, is zero. Then integrating this equation again gives

$$x = \left(\frac{F}{M}\right) \frac{t^2}{2} + x(0)$$

$$= \left(\frac{F}{M}\right) \frac{t^2}{2} \qquad (1.31)$$

since $x(0) = 0$. Thus the distance d is travelled in t_d s such that

$$t_d = \sqrt{\left(\frac{2Md}{F}\right)} . \qquad (1.32)$$

The velocity of the mass at that instant may be written

$$\dot{x}(t_d) = \frac{F}{M} \sqrt{\left(\frac{2Md}{F}\right)}$$

$$= \sqrt{\left(\frac{2Fd}{M}\right)} . \qquad (1.33)$$

The energy gained by the mass from the work done on it is

$$E_k = Fd$$

$$= \tfrac{1}{2} M [\dot{x}(t_d)]^2 \qquad (1.34)$$

and it is contained in the mass in the form of so-called kinetic energy, energy possessed by virtue of the motion of the mass. Kinetic energy is measured in units of joules (J) the same as for mechanical work. In general, a body moving with velocity $\dot{x}$ has a kinetic energy of $\tfrac{1}{2} M \dot{x}^2$. The analogous expression for the kinetic energy of a rotating body is $\tfrac{1}{2} I \dot{\theta}^2$ where I is the mass moment of inertia of the body about the axis of rotation.

If a force barely greater than the weight of a mass acts vertically upwards

FUNDAMENTAL MECHANICAL QUANTITIES

and lifts the mass a height d m, the energy gained by the mass is again equal to the work done thus

$$E_p = Fd$$
$$= Mgd \qquad (1.35)$$

where g is the acceleration due to gravity ($= 9.81$ m/s^2) and the energy gained is called potential energy by virtue of the increase in height of the object in the Earth's gravitational field.

Note that if a mass has this latter amount of potential energy and is allowed to fall under gravity a distance d, it is accelerated to a velocity $\dot{x}_d$ such that

$$\dot{x}_d = \sqrt{(2gd)} \qquad (1.36)$$

from equations 1.33 and 1.35. In this case the loss of potential energy in falling the distance d is equal to the gain of kinetic energy. The total mechanical energy is said to be conserved. This is an elementary example of the principle of conservation of mechanical energy.

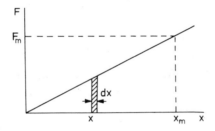

Figure 1.13 A conventional force F versus deflection x graph for a spring, showing the work done (the cross-hatched area) in stretching the spring from x to $x + dx$.

Work must be done in order to deform a spring and therefore energy may be stored in a spring. This form of stored mechanical energy is also termed potential energy. To calculate the amount of energy stored in a spring note from figure 1.13 that the element of work done in stretching the spring from x to $x + dx$, during which elemental interval the force is effectively constant at $F = Cx$, is

$$d(Wk) = F\,dx$$
$$= Cx\,dx \qquad (1.37)$$

Therefore the total work done in stretching the spring from 0 to x_m may be found by integrating both sides of this equation.

$$\int d(Wk) = Wk$$

$$= C \int_0^{x_m} x \, dx$$

$$= \tfrac{1}{2} C x_m^2 \tag{1.38}$$

$$= \tfrac{1}{2} F_m x_m . \tag{1.39}$$

The work done and hence the potential energy stored is the same whether the spring is stretched ($+x$) or compressed ($-x$) as is clear from equation 1.38. In each case the energy stored is the area under the F/x graph up to the maximum deformation. This energy storage relationship holds for any type of spring. When a spring is deflected it contains potential energy.

One further parameter relating to work and energy should be noted. That is, the power P, which is the rate at which work is done, thus

$$P = \dot{W}k . \tag{1.40}$$

The unit of power is the watt (W) which is equivalent to joule per second (J/s). Clearly, also, the rate of transfer of energy is power.

Example 1.4

Calculate the energies stored in the two springs of example 1.3.

Solution

(a) Tension

$$E_p = \tfrac{1}{2} F \Delta L = 0.5 \times 10^3 \times 50 \times 10^{-6} = 0.025 \text{ J} .$$

(b) Three-point-bending

$$E_p = \tfrac{1}{2} F \delta = 0.5 \times 10^3 \times 0.25 = 125 \text{ J} .$$

FUNDAMENTAL MECHANICAL QUANTITIES

1.5 Units

SI (Système Internationale) units are used throughout this text. The basic SI units needed are those of table 1.1.

The derived SI units encountered with some frequency in the text are listed in table 1.2.

Table 1.1 Basic SI Units

Quantity	Unit	Unit Symbol
mass	kilogram	kg
length	metre	m
time	second	s
angle	radian	rad
temperature	degrees kelvin	K
current	ampere	A

Table 1.2 Derived SI Mechanical Units

Quantity	Unit	Unit Symbol
angle	degree	°(rad × 180/π)
velocity	metre/second	m/s
acceleration	metre/second/second	m/s^2
force	newton	N (= kg m/s^2)
pressure or stress	pascal	Pa (= N/m^2)
momentum	kilogram metre/second	kg m/s
work or energy	joule	J (= N m)
power	watt	W (= J/s)
impedance	—	kg/s
frequency	hertz	Hz (= 1/s)
angular velocity	radian/second	rad/s
angular acceleration	radian/second/second	rad/s^2
torque	newton metre	N m
temperature	degree Celcius	°C (K+273)

1.6 Experiments

(a) Measurement of Mass

The mass of an object is measured using a laboratory balance. The technique is to compare the unknown mass with a known mass. The unknown mass is placed on one pan of the balance and known masses are added to the second pan until balance is obtained. At balance, the torque exerted by the weight of the unknown mass about the fulcrum is equal to that exerted by the known mass. The lever arms on both sides, however, are of equal length and so the two masses must be equal.

Larger masses may be compared with a sturdy lever. A metre stick with a hole at the centre of its length and a pin through this hole as fulcrum can serve as the lever. The known mass is hung at one end of the lever and the unknown mass may be hung at various distances along the lever arm until balance is found. By calculating the two equal torques the unknown mass may be found.

(b) To Verify Newton's Second Law using Fletcher's Trolley

Fletcher's trolley runs on freely moving wheels and its mass may be readily changed by adding or removing known masses to prepared cavities in it. A horizontal string from the trolley passes over a pulley, and a pan, in which different masses may be placed, is attached to the other end of the string. The weight of the mass in the pan is the accelerating force on the mass of the trolley. The acceleration of the trolley may be calculated from measurement of the times taken for the trolley starting from rest to pass fixed marks at measured intervals along the horizontal table. Alternatively, a pen attached to a vibrating arm, oscillating at constant period, can be made to write on a sheet of paper attached to the top of the trolley. The distance travelled is plotted against half of the square of the time from the start to reach the distance and (see problem 1.1) the slope of this graph is the acceleration. This experiment can be used to show that the acceleration is directly proportional to the applied force and inversely proportional to the mass of the trolley.

(c) To Verify the Vector Addition of Forces

If three known masses are all attached to strings, the other ends of which are all tied together at a point, and two of the strings are passed over two pulleys and the third allowed to hang freely, an equilibrium will be established with the strings radiating from the common point at various angles. These angles may be measured and the triangle of forces, figure 1.2, drawn for the three tensile forces in the strings. One can check that the angles of the triangle correspond to those measured or else, by constructing the triangle with two of the forces and the three angles measured, the third side of the triangle should equal the third force. Different combinations of masses and hence weights may be used to develop a thorough verification.

FUNDAMENTAL MECHANICAL QUANTITIES

(d) To Determine the Moment of Inertia of a Flywheel

A wheel plus axle of radius r and total mass M, is allowed to roll on the axle down an inclined plane of angle ϑ. In dropping through a vertical height h, it travels a distance d along the inclined plane such that $h/d = \sin \vartheta$. The time t taken for the wheel to travel the distance d is measured. The initial potential energy is converted into kinetic energy of rotation and of linear translation. In consequence the following equation relates the two measured quantities t and ϑ.

$$t^2 = \frac{2d(M + I/r^2)}{Mg} \frac{1}{\sin \vartheta}.$$

ϑ is varied and t measured for each ϑ. The graph of t^2 against $1/\sin \vartheta$ yields a slope from which the moment of inertia I can be calculated.

(e) Angular Acceleration of the Flywheel by Torques

A torque may be applied to the flywheel, mounted in bearings, by wrapping a string around the axle and tying a known mass to the free end of the string. If the mass drops through a measured distance s before the string disengages, the loss of potential energy of the mass must equal the rotational kinetic energy gained by the flywheel plus any energy used to overcome friction, together with the gain of kinetic energy of the mass itself. If the friction be neglected and both the time t for the drop of the mass through the distance s and the total number of rotations of the wheel from the start until t are measured, then the angular acceleration can be written

$$\ddot{\vartheta} = \frac{Mgr}{I}$$

from which the angular displacement is given by

$$\vartheta = \tfrac{1}{2} (\ddot{\vartheta}) t^2 .$$

If a number of values of M be used, then a graph of ϑ versus Mt^2 will be a straight line through the origin with a slope $\tfrac{1}{2} gr/I$.

(f) To Verify Hooke's Law with a Vertical Spiral Spring

The spring is mounted with the top end rigidly fixed and a pan to receive known masses attached to the bottom end. A horizontal pointer is attached to the pan indicating the location of the pan on a vertically mounted metre stick. The mass in the pan is gradually increased from zero and the reading of the pointer on the metre stick read at each value of mass. A graph of the extension of the spring versus the weight of the mass in the pan produces a straight line as predicted by Hooke's law. The slope of the graph is the inverse of the spring constant C.

(g) Measurement of Young's Modulus of a Wire

Two lengths of identical wire are hung vertically downwards, one a reference wire and the other the test piece. A fixed weight is hung on the end of the former while the weight on the test wire is gradually built up from zero. The reference wire has a scale attached and the test wire has a vernier attached. As the force on the test wire is built up, the vernier is read. Each change in vernier reading from the initial reading at zero load is divided by the length of the wire from the vernier mount to the top mounting to yield the strain. The diameter of the wire is first read and the cross-sectional area calculated. Each force setting is then converted to a stress. The stress is plotted versus the corresponding strain and the slope of this straightline graph is the Young's modulus of the material of the wire.

If a universal testing apparatus is available, which can be used to apply high forces to specimens of materials, larger cylindrical specimens can be tested but the basic approach is the same.

Problems

1.1 Two equal forces of 100 N act in the horizontal plane on the same point. One is directed at 60° to the x-axis while the other is at 30° to this axis on the other side. Find the resultant force in magnitude and direction both graphically and analytically.

1.2 An object of mass 55 kg sits on a horizontal table. Two strings are attached to the centre of gravity of the object and exert tensile forces on it in directions ϑ and 2ϑ to the vertical and on either side of the vertical. One of these forces is 1.3 times the other and the table exerts an upward reaction force of one half the weight of the object. Find the two forces and their directions.

1.3 A flywheel of radius 2 m has a rope attached at 0.5 m from the rim and a force of 450 N exerted by this force along the direction parallel to the tangent to the rim at that point but the rope makes an angle of 35° with the plane of the wheel. What braking force at the rim keeps the wheel motionless?

1.4 An object of weight 10 N is on an inclined plane 25° to the horizontal. A string attached to the object exerts a constant force of 8 N but its angle to the vertical can be varied from 0° to 65°. Plot a graph of the net force on the object as a function of this angle. At what angle is the net force along the plane zero?

1.5 A body of mass 5 kg hangs on the bottom of a string of length 1 m. What horizontal force must be exerted on the mid-point of the string to deflect it by 20 cm and what work is done in this manoeuvre?

1.6 A force of 1000 N acts on a vehicle for 10 s, increasing the velocity from rest to 10 m/s. What is the mass of the vehicle and how far does it travel in the 10 s?

1.7 A rocket of mass 10 kg is accelerated vertically upwards at a steady 50 m/s^2 for 40 s. What is the thrust or force of the engine, what is the height after the 40 s and what is the velocity then?

1.8 A ball is thrown horizontally at a velocity of 10 m/s at 6 m above the ground. How long does it take for the ball to strike the ground and how far horizontally does it travel before it does so? Draw a diagram of the trajectory of the ball.

1.9 A flywheel of moment of inertia 1500 kg m^2 is rotating at 1000 revolutions per minute when a steady braking force is applied at the rim at 1 m radius. The wheel comes to rest in 5 s. Find the braking force and the number of rotations during the braking phase.

1.10 An object of mass 50 kg falls from rest at a height of 10 m on to a spring of height 1 m. The maximum deflection of the spring is 1 cm. Find the spring constant.

1.11 A railway wagon of mass 20 Mg moving along a horizontal track is halted by a buffer spring of constant 150 kN/m which is shortened by 0.1 m in the process. What was the impact velocity of the wagon?

1.12 A body of mass 10 kg sits on the top of a spring of a spring of length 0.1 m which is compressed by 10 per cent. The body is then released and reaches a maximum height of 2 m. Find the spring constant.

1.13 A bar of length 2 m has a cross-sectional area of 1 cm^2. It is rigidly fixed at the top in a vertical position. At the lower end there is rigidly attached a flange. A loose-fitting concentric ring of mass 1000 kg is dropped from the 2 m height and causes the flange to be deflected downwards a maximum of 1 mm. What is the Young's modulus value for the material of the bar? Assume no loss mechanisms.

1.14 A waterfall has a steady flow of 10^5 m^3 each 24 hour day and the height of the fall is 50 m. What is the power expended by gravity on the water?

1.15 A beam of rectangular cross-section 1 cm × 2 cm and length 2 m is mounted as a cantilever. When a mass of 10 kg is hung from the free end the vertical deflection of the end is 5 cm. Find the value of Young's modulus for the material of the beam. If the beam is mounted with its width, instead of its thickness, in the vertical direction what would the deflection be?

1.16 An ice skater, viewed as a vertical cylinder of mass 250 kg and diameter 0.4 m, is spinning at 4 rotations each second. What will be the rotational speed when the arms, viewed as point masses 15 kg each on a massless rod of length 0.5 m, are extended radially outwards?

1.17 A flywheel of moment of inertia 400 kg m^2 is steadily accelerated from rest to 1000 revolutions per minute in 15 s. Find the torque and the final kinetic energy.

Further Reading

Bishop, R.E.D., *Vibration* (Cambridge University Press, 1965)
Daish, C.B. and Fender, D.H., *Experimental Physics,* 2nd edition (Hodder & Stoughton, London, 1976)
MacDonald, S.G.G. and Burns, D.M., *Physics for the Life and Health Sciences* (Addison-Wesley, Reading, MA., 1975)
Ryder, G.H., *Strength of Materials,* 3rd edition (Macmillan, London, 1973)
Sears, F.W. and Zemansky, M.W., *University Physics*, 4th edition (Addison-Wesley, Reading, MA., 1972)
Steidel, R.F. Jr., *An Introduction to Mechanical Vibrations,* 2nd edition (Wiley, New York, 1979)
Tyler, F., *A Laboratory Manual of Physics*, 5th edition, SI version (Arnold, London, 1977)

2 Free Harmonic Oscillator

2.1 Introduction and Objectives

In this chapter the aim is to describe the oscillatory behaviour of a simple system when vibrating freely, that is, without any external influences after the initial start-up.

After reading this chapter, the student should be able to

(a) develop the equation of motion for a simple harmonic oscillator;
(b) describe graphically the motion of the mass;
(c) describe the forms of energy in the system;
(d) use a variety of initial conditions to help specify the oscillatory behaviour of the system;
(e) outline a number of simple harmonic oscillators;
(f) develop electrical analogues of such oscillators.

2.2 The Simple Harmonic Oscillator

The simplest type of lumped-parameter vibrating mechanical system is the harmonic oscillator or mass/spring system shown in figure 2.1 which has only one degree of freedom — along the horizontal x-axis. At rest, the mass M is stationary and the spring of constant C is unstrained. No force then acts on the system along the horizontal direction and there is no mechanical energy in the system. Assume a perfectly smooth frictionless table on which the mass rests.

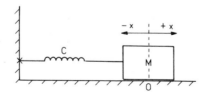

Figure 2.1 A simple harmonic oscillator with one degree of freedom in the x-direction. It consists of a mass M attached to a spring of constant C. The other end of the spring is rigidly fixed.

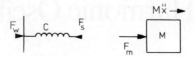

Figure 2.2 The free-body diagrams of the spring under the two equal and opposite forces F_s and F_w and of the mass under the force F_m which accelerates the mass by $\ddot{x}$.

If the mass be moved to the left a distance x, the free-body diagram for the system becomes that of figure 2.2. Here the force on the spring F_s, using Hooke's law of equation 1.24, may be expressed thus

$$F_s = Cx \ . \tag{2.1}$$

The force on the mass, F_m, which acts to accelerate the mass, must be equal and opposite to the spring force and so is given by

$$\begin{aligned} F_m &= -F_s \\ &= -Cx \\ &= M\ddot{x} \ , \end{aligned} \tag{2.2}$$

and so

$$\ddot{x} + \frac{C}{M}x = 0 \ . \tag{2.3}$$

This is the equation of simple harmonic motion (S.H.M.), the acceleration of the mass being directly proportional to the negative of the displacement. The most general solution to this equation may be shown by substitution in the equation to take the form

$$x = \acute{A} \, e^{-j(\omega_0 t + \varphi)} + \acute{B} \, e^{j(\omega_0 t + \psi)} \tag{2.4}$$

where $\acute{A}$, $\acute{B}$, φ and ψ are constants and

$$\omega_0 = \sqrt{\left(\frac{C}{M}\right)} \ , \tag{2.5}$$

is the so-called natural angular frequency of the system. The solution of equation 2.4 is complex but the actual solution in the case of a physical oscillator can only be real and so the true solution will be

$$x = \text{Re}[\acute{A} \, e^{-j(\omega_0 t + \varphi)}] + \text{Re}[\acute{B} \, e^{j(\omega_0 t + \psi)}] \ . \tag{2.6}$$

FREE HARMONIC OSCILLATOR

But

$$e^{-j(\omega_0 t + \varphi)} = \cos(\omega_0 t + \varphi) - j \sin(\omega_0 t + \varphi) \tag{2.7}$$

and

$$e^{j(\omega_0 t + \psi)} = \cos(\omega_0 t + \psi) + j \sin(\omega_0 t + \psi) . \tag{2.8}$$

Therefore equation 2.6 may be rewritten

$$x = \acute{A} \cos(\omega_0 t + \varphi) + \acute{B} \cos(\omega_0 t + \psi) \tag{2.9}$$

which can be recast into

$$x = A \cos(\omega_0 t + \gamma) \tag{2.10}$$

in which

$$A = \sqrt{[\acute{A}^2 + \acute{B}^2 + 2\acute{A}\acute{B} \cos(\varphi - \psi)]} \text{ and } \tan \gamma = \frac{\sin \varphi + (\acute{B}/\acute{A}) \sin \psi}{\cos \varphi + (\acute{B}/\acute{A}) \cos \psi} \tag{2.11}$$

A is the amplitude of the oscillatory displacement, the maximum excursion of the mass on either side of the resting location. γ is the phase angle that relates the starting conditions of the motion to the sine wave. Both the amplitude and phase angle are determined by the initial conditions (those at $t = 0$) of the motion. They are independent of the system components.

The displacement described by equation 2.10 is sinusoidal as shown in figure 2.3.

The natural frequency of this oscillatory motion f_0, measured in hertz (Hz) is given by

$$f_0 = \frac{\omega_0}{2\pi}$$

$$= \frac{1}{2\pi} \sqrt{\left(\frac{C}{M}\right)} . \tag{2.12}$$

The period of the oscillation T, measured in seconds (s), is

$$T = \frac{1}{f_0}$$

$$= 2\pi \sqrt{\left(\frac{M}{C}\right)} . \tag{2.13}$$

Thus the frequency and the period are determined directly by the parameters of the system and are independent of the initial conditions.

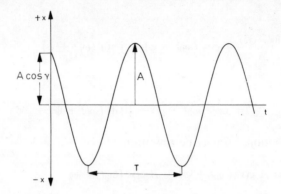

Figure 2.3 The displacement of the mass as a function of time. The amplitude A of the oscillation and the period T are shown. Also the initial value of the displacement helps to determine the value of the phase angle γ.

Example 2.1

A system such as that of figure 2.1 has a period of oscillation of 6 s. The mass is 10 kg. Find the frequency and the spring constant. If the mass is doubled what will the period be?

Solution

Here $T_1 = 6$.

Therefore $f_{01} = \dfrac{1}{T_1} = 0.17$ Hz .

But from equation 2.12

$$\frac{C}{M_1} = (2\pi f_{01})^2 = 1.14 \text{ and so}$$

$$C = 11.4 \text{ N/m} .$$

If $M_2 = 20$ kg, equation 2.13 allows one to write

$$T_2 = 2\pi \sqrt{\left(\frac{M_2}{C}\right)} = 8.5 \text{ s} .$$

FREE HARMONIC OSCILLATOR

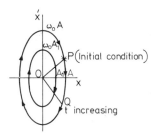

Figure 2.4 The velocity $\dot{x}$ versus displacement x plots for the simple harmonic motion of the mass in the oscillator. The loops are ellipses of axes x (max) or A and $\dot{x}$(max) or $\omega_0 A$. The radial line OQ rotates clockwise at ω_0 rad/s.

2.3 Graphical Display of Simple Harmonic Motion

Since the displacement of the mass can be described by equation 2.10

$$x = A \cos(\omega_0 t + \gamma)$$

the velocity may be derived by differentiation, thus

$$\dot{x} = -\omega_0 A \sin(\omega_0 t + \gamma) = \omega_0 A \cos\left(\omega_0 t + \gamma + \frac{\pi}{2}\right). \quad (2.14)$$

The velocity is therefore sinusoidal with the same frequency f_0 as the displacement but is $\pi/2$ rad ahead of the displacement.

This oscillatory motion may be displayed in the $\dot{x}/x$ plane, the so-called phase plane, as in figure 2.4. This is a plot of the velocity against the displacement at each instant during the motion. The display is an elliptical loop, starting at the initial condition point P, which is completed once each period T. The ellipse is traced out by a variable length radial line OQ, rotating clockwise about the origin of axes 0, at an angular velocity ω_0 rad/s.

There are four points of special note about the elliptical loop in the phase plane:

(a) the ellipse passes through the initial condition point;
(b) the horizontal or x-axis of the ellipse is equal to $2A$, the peak to peak displacement of the mass;
(c) the vertical or $\dot{x}$-axis of the ellipse is equal to $2\omega_0 A$, the peak-to-peak velocity of the mass;
(d) the loop is traversed in the clockwise direction repetitively, at a frequency f_0 Hz.

2.4 Forces in the Oscillator

The acceleration of the mass is found by differentiating the velocity of equation 2.14.

$$\ddot{x} = -\omega_0^2 A \cos(\omega_0 t + \gamma)$$
$$= -\omega_0^2 x \ . \tag{2.15}$$

This equation is another more general version of equation 2.3 for a system which executes simple harmonic motion. In such a system the constant of proportionality between the acceleration and the negative of the displacement is the square of the angular velocity of the radial line OQ in the $\dot{x}/x$ plot of figure 2.4.

From equation 2.2 the force exerted by the spring on the mass is

$$F_m = -M\omega_0^2 A \cos(\omega_0 t + \gamma) \ , \tag{2.16}$$

an oscillatory force of frequency f_0 and of amplitude $M\omega_0^2 A$ or CA. This force is equal and opposite to the force F_s exerted on the spring by the mass. F_s in turn is equal to the force exerted by the left hand end of the spring on the rigid upright wall in figure 2.1.

2.5 Energy in the System

The mechanical energy in the harmonic oscillator falls into the two categories, the kinetic energy of the mass E_k and the potential energy of the spring E_p. The kinetic energy of the mass may be written from equations 1.34 and 2.14,

$$E_k = \tfrac{1}{2} M\dot{x}^2$$
$$= \tfrac{1}{2} M\omega_0^2 A^2 \sin^2(\omega_0 t + \gamma) \ . \tag{2.17}$$

The potential energy stored in the spring may be written, using equations 1.38 and 2.10

$$E_p = \tfrac{1}{2} Cx^2$$
$$= \tfrac{1}{2} CA^2 \cos^2(\omega_0 t + \gamma) \tag{2.18}$$

FREE HARMONIC OSCILLATOR

Therefore the total energy in the system E_t is the sum of these two categories thus

$$E_t = E_k + E_p$$
$$= \tfrac{1}{2} A^2 [M\omega_0^2 \sin^2(\omega_0 t + \gamma) + C \cos^2(\omega_0 t + \gamma)]. \quad (2.19)$$

But from equation 2.5, $M\omega_0^2 = C$, and so

$$E_t = \tfrac{1}{2} A^2 C$$
$$= \tfrac{1}{2} A^2 M\omega_0^2. \quad (2.20)$$

The three energies E_t, E_k and E_p are plotted in figure 2.5. The total mechanical energy in the system E_t is constant, that is, it is conserved but the form of that energy oscillates between kinetic and potential forms.

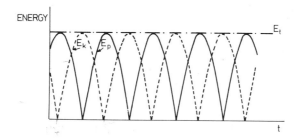

Figure 2.5 The variation of kinetic energy E_k and potential energy E_p in the harmonic oscillator with time. The total energy E_t is a constant illustrating the principle of conservation of mechanical energy.

At certain instants all of the energy is in the kinetic form, that is, when

$$t = \frac{n\pi}{2\omega_0} - \frac{\gamma}{\omega_0} \quad (2.21)$$

for all positive integer values of n. These are the instants at which the mass is moving at maximum velocity through the rest position, $x = 0$, in either direction. At that position, the spring is at its resting length and therefore possesses no potential energy.

At other instants given by

$$t = \frac{(n-1)\pi}{\omega_0} - \frac{\gamma}{\omega_0} \quad (2.22)$$

for all positive integer values of n, all of the energy is in the potential form.

At these instants, the mass is at one or other extreme point of the displacement and is momentarily motionless. It then has no kinetic energy and only the spring contains the stored potential energy.

2.6 Initial Conditions

In equation 2.10, describing the motion of the mass in the harmonic oscillator, there are two parameters, the displacement amplitude A and the phase angle γ which are determined by the state of the system at $t = 0$, the so-called initial conditions. Thus at $t = 0$, the initial displacement is

$$x(0) = A \cos \gamma, \qquad (a)$$

the initial velocity is

$$x'(0) = -\omega_0 A \sin \gamma, \qquad (b)$$

and the initial acceleration is

$$x''(0) = -\omega_0^2 A \cos \gamma. \qquad (c)$$

The value of the initial force on the mass is

$$F_m(0) = -M\omega_0^2 A \cos \gamma, \qquad (d)$$

while the initial force on the wall is $\qquad$ (2.23)

$$F_s(0) = M\omega_0^2 A \cos \gamma$$
$$= CA \cos \gamma. \qquad (e)$$

The initial kinetic energy of the mass is

$$E_k(0) = \tfrac{1}{2} M\omega_0^2 A^2 \sin^2 \gamma, \qquad (f)$$

the initial potential energy stored in the spring is

$$E_p(0) = \tfrac{1}{2} CA^2 \cos^2 \gamma \qquad (g)$$

FREE HARMONIC OSCILLATOR

while the total initial energy in the system is

$$E_t(0) = \tfrac{1}{2}CA^2 \ . \qquad \text{(h)}$$

Clearly not all of these initial conditions are independent. Thus for instance

$$\begin{aligned}E_p(0) &= \tfrac{1}{2}C\,[x(0)]^2 \\ &= \tfrac{1}{2}\frac{M}{\omega_0^2}\,[\ddot{x}(0)]^2 \\ &= \tfrac{1}{2}\frac{[F_m(0)]^2}{C}\ . \end{aligned} \qquad (2.24)$$

Also

$$E_k(0) = \tfrac{1}{2}M[\dot{x}(0)]^2 \ . \qquad (2.25)$$

Knowledge of any two independent initial conditions, one involving $\cos \gamma$, either $x(0)$, $\ddot{x}(0)$, $F_m(0)$, $f_s(0)$ or $E_p(0)$, and one involving $\sin \gamma$, either $\dot{x}(0)$ or $E_k(0)$, or else one out of these groups together with E_t, allows calculation of A and γ and hence the complete specification of the sinusoidal motion.

As an example, take the following initial conditions, the displacement p and the velocity q at $t = 0$

$$x(0) = p \text{ and } \dot{x}(0) = q \ . \qquad (2.26)$$

From equations 2.23(a) and (b), $A \cos \gamma = p$ and $-\omega_0 A \sin \gamma = q$. Thus

$$A \sin \gamma = -\frac{q}{\omega_0} \text{ and}$$
$$A^2 = p^2 + q^2/\omega_0^2 \ .$$

Therefore

$$A = \sqrt{(p^2 + q^2/\omega_0^2)} \ . \qquad (2.27)$$

Also

$$\tan \gamma = -\frac{q}{\omega_0 p}$$

and so

$$\gamma = \tan^{-1}\left(-\frac{q}{\omega_0 p}\right) . \qquad (2.28)$$

Example 2.4

The oscillator of figure 2.1, in which the mass is in motion in the horizontal direction, has a spring of constant 20 N/m. At a particular instant the mass has a velocity of +1 m/s while the spring is stretched by +10 cm. It takes 3 s for the mass to again reach the same phase of the motion. Find the magnitude of the mass, the total mechanical energy in the system and the amplitude of the vibration.

Solution

The period $T = 3$ s. Therefore

$$\omega_0 = \frac{2\pi}{T} = 2.09 \text{ rad/s}$$

and

$$M = C/\omega_0^2 = 4.56 \text{ kg} .$$

At one instant $\dot{x} = 1$ m/s and $x = 0.1$ m. From equation 2.17, the kinetic energy of the mass then is

$$E_k = \tfrac{1}{2} M \dot{x}^2 = 2.28 \text{ J} .$$

From equation 2.18 the potential energy stored in the spring is then

$$E_p = \tfrac{1}{2} C x^2 = 0.1 \text{ J} .$$

So the total energy in the system is the sum of these two

$$E_t = E_k + E_p = 2.38 \text{ J} .$$

This total energy is conserved so that when $x = A$, $E_k = 0$ and $E_p = E_t = \tfrac{1}{2} C A^2$. Therefore

$$A = \sqrt{(2E_t/C)} = 0.49 \text{ m, the amplitude of the vibration.}$$

2.7 Other Simple Harmonic Oscillators

Many types of discrete component simple harmonic oscillators, other than that of figure 2.1, are physically realisable. One familiar example is the simple pendulum of figure 2.6, which consists of a massless string of length L with one end fixed at point P and the other end attached to the point

FREE HARMONIC OSCILLATOR

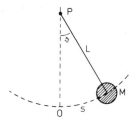

Figure 2.6 A simple pendulum, consisting of mass M attached to a string of length L the other end of which is fixed at P. When the string is deflected by ϑ from the vertical the mass is displaced along the circumference of the circle of radius L by s.

mass M. When the string is deflected by an angle ϑ from the vertical, the free-body diagram for the mass is that shown in figure 2.7(a). The string exerts a force F on the mass while Mg is the weight of the mass acting vertically downwards. The weight force may be decomposed into two components $Mg \cos \vartheta$ acting along the direction of the string and hence the direction of F and $Mg \sin \vartheta$ perpendicular to the direction of the string. This latter force is unbalanced and therefore accelerates the mass along the arc of the circle, centre at P and of radius L, towards the point O, vertically below the point of support P. In this oscillator the restoring force is due to gravity, in contrast with the system of figure 2.1, in which it is due to the spring force.

The motion of the mass is along the arc of the circle. Consider in this case displacement from O along the arc to be s. Of course

$$s = L\vartheta \text{ and } \ddot{s} = L\ddot{\vartheta} . \tag{2.29}$$

Figure 2.7 The free-body diagram for the mass in the simple pendulum (a) in the actual circumstance when the string is deflected from the vertical and (b) with the weight of the mass decomposed into two components, one along the direction of the string and one perpendicular to the string.

The unbalanced force causes the acceleration of the mass, thus

$$-Mg \sin \vartheta = M\ddot{s}$$
$$= ML\ddot{\vartheta}, \qquad (2.30)$$

or, after simplification

$$\ddot{\vartheta} + \frac{g}{L} \sin \vartheta = 0 . \qquad (2.31)$$

If the deflection of the mass from O is kept small, ϑ is small and $\sin \vartheta = \vartheta$. Then equation 2.31 reduces to

$$\ddot{\vartheta} + \frac{g}{L} \vartheta = 0 \qquad (2.32)$$

or

$$\ddot{s} + \frac{g}{L} s = 0 , \qquad (2.33)$$

which is identical in form with equation 2.3, where now

$$\omega_0 = \sqrt{\left(\frac{g}{L}\right)} \qquad (2.34)$$

and

$$T = 2\pi \sqrt{\left(\frac{L}{g}\right)} . \qquad (2.35)$$

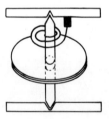

Figure 2.8 A rotational simple harmonic oscillator.

The rotational vibratory system shown in figure 2.8 is also a simple harmonic motion system. Take the disc and rod to have a mass moment of inertia I about the rotational axis and the torsion spring to have a torsional stiffness value C. When the disc is twisted an angle ϑ from the resting position, an unbalanced restoring torque Tr, comes into play where

FREE HARMONIC OSCILLATOR

$$\text{Tr} = C\vartheta \ . \tag{2.36}$$

This causes a rotational acceleration of the rod and disc, thus

$$-C\vartheta = I\vartheta'' \tag{2.37}$$

or

$$\vartheta'' + \frac{C}{I}\vartheta = 0 \ . \tag{2.38}$$

Simple harmonic rotational motion results, with

$$\omega_0 = \sqrt{\left(\frac{C}{I}\right)} \tag{2.39}$$

and

$$T = 2\pi\sqrt{\left(\frac{I}{C}\right)} \ . \tag{2.40}$$

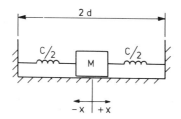

Figure 2.9 A harmonic oscillator consisting of a single mass M attached in series with two identical springs of constant $C/2$. The other ends of the springs are rigidly fixed.

The discrete component system of figure 2.9 may also be shown to execute simple harmonic motion. This system consists of a mass M attached to two identical springs of original length l_0, and spring constant $C/2$. If the mass be displaced to the right by $+x$, the free-body diagram of figure 2.10 results. Therefore the net force on the mass accelerates it thus

$$-\frac{C}{2}[(d - l_0 + x) - (d - l_0 - x)] = M x''$$
$$= -Cx \ . \tag{2.41}$$

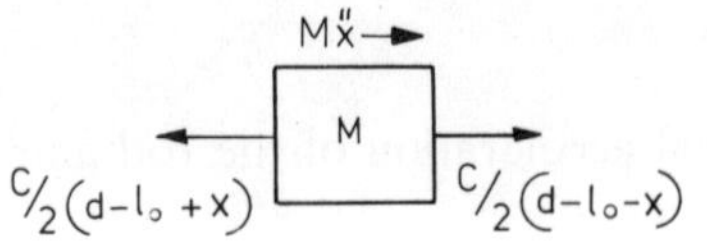

```
                    Mẍ →
         ┌─────────────┐
────     │      M      │     ────
 C/2(d-l₀+x)           C/2(d-l₀-x)
```

Figure 2.10 The free-body diagram for the mass in figure 2.9 indicating the two spring forces producing an acceleration $\ddot{x}$.

That is

$$\ddot{x} + \frac{C}{M}x = 0$$

which is equation 2.3. Clearly this mass behaves identically with that in the system of figure 2.1.

Example 2.5

A simple pendulum consists of a mass of 10 kg attached to a string of length 10 m. Initially the mass is displaced 1 m from the vertical line through the point of support of the string and released from rest. Find the period of the oscillation and the total energy in the system. Write down the equation for the angular motion of the string. Take $g = 9.8$ m/s².

Solution

From equation 2.35, the period T is given by

$$T = 2\pi\sqrt{(10/9.8)} = 6.35 \text{ s}.$$

Therefore

$$\omega_0 = \sqrt{(9.8/10)} = 0.99 \text{ rad/s}.$$

The total energy is the amount of initial potential energy of the mass referred to the lowest point of the motion

$$E_p(0) = Mg\,[10 - 10\cos\vartheta(0)]$$

But at $t = 0$, $\sin\vartheta(0) = 1/10$ and $\vartheta(0) = 0.1$ rad. So

$$E_p(0) = 98\,(10 - 9.9) = 9.8 \text{ J}.$$

FREE HARMONIC OSCILLATOR

Finally the angular motion of the string is

$$\vartheta = \vartheta(0) \cos \omega_0 t = 0.1 \cos 0.99 t \text{ rad} .$$

2.8 Electrical Analogues of the Simple Harmonic Oscillator

Consider the parallel LC electrical circuit of figure 2.11. The voltage rise v_c, across the capacitance C is equal to but of opposite polarity to the voltage drop v_1 across the inductance L.

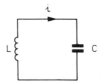

Figure 2.11 An inductor/capacitor circuit which is analogous to a simple mechanical harmonic oscillator.

$$\begin{aligned} v_c &= \frac{q}{C} \\ &= -v_1 \\ &= -L\,\dot{i} \end{aligned} \quad (2.42)$$

where q is the charge on the capacitor and i is the current flowing at any instant. But

$$i = \dot{q} \quad (2.43)$$

and so equation 2.42 becomes

$$L\ddot{q} + \frac{1}{C}q = 0$$

or

$$\ddot{q} + \frac{1}{LC}q = 0 . \quad (2.44)$$

This equation is identical in form with equation 2.3 except that q takes the place of x, L that of M and $1/C$ that of C (spring constant).

Table 2.1 The Analogous Quantities in Mechanical and Electrical Systems

Mechanical Quantity	Electrical Quantity
displacement, x	charge quantity, q
velocity, $\dot{x}$	current, i
acceleration, $\ddot{x}$	rate of change of current, $\dot{i}$
mass, M	inductance, L
spring constant, C	inverse of capacitance, $1/C$
force, F	potential difference, v
damping coefficient, b	resistance, R

Thus the parallel LC circuit is directly analogous to the discrete component simple harmonic oscillator of figure 2.1 and the range of analogous quantities and parameters are listed in table 2.1. The period T of the electrical oscillations may be written from equation 2.13 by substituting analogous parameters from the table, thus

$$T = 2\pi\sqrt{(LC)} \ . \tag{2.45}$$

The complete description of the oscillations of q is given from equation 2.10

$$q = A \cos(\omega_0 t + \gamma) \tag{2.46}$$

where

$$\omega_0 = \sqrt{\frac{1}{(LC)}} \ , \tag{2.47}$$

A is the amplitude of the oscillation and γ is a phase angle relating to the initial conditions in the circuit.

Wide ranges of values for the electrical circuit elements can be readily chosen and the behaviour of such circuits rapidly tested on the bench using an oscilloscope and other common electrical test equipment. Electrical analogue circuits therefore offer the possibility of predicting the behaviour of a mechanical system in the aftermath of various initial conditions. The electrical system parameters can be modified to achieve desired behaviour and therefore allow the mechanical system to be designed for specific behaviour.

FREE HARMONIC OSCILLATOR

Analogue computer techniques allow still more accurate and flexible simulation of a mechanical oscillatory system. In an analogue computer the central active components are high-gain operational amplifiers, connected to function as differentiators, integrators, summing amplifiers and inverting amplifiers. Figure 2.12 shows an analogue computer scheme to simulate the mechanical system of figure 2.1 and to solve equation 2.3 for that system. The simulation consists of two integrators and one inverting amplifier. In operation, the switches are connected as shown. The first integrator 1 receives an input voltage $\ddot{x}$ and produces an output voltage $-\dot{x}$. This latter is the input to the second integrator 2 which has an output voltage x. If C/M is less than unity, a potential divider produces a voltage of C/Mx as input to the unity gain inverting amplifier 3. If C/M is greater than unity, the potential divider is omitted and the inverting amplifier is designed to have a gain of $-C/M$.

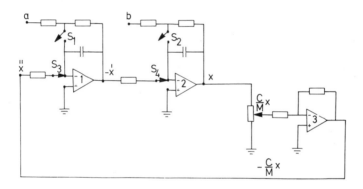

Figure 2.12 An analogue computer set-up which may be used to solve the equation of motion, equation 2.3, of the harmonic oscillator. (After Clayton).

Initial conditions, such as the initial values of velocity $\dot{x}$, or displacement x, may be set by charging the feedback capacitors of integrators 1 and 2 respectively to appropriate voltage levels $\dot{x}(0)$ and $x(0)$. This charging manoeuvre is done with switches S_3 and S_4 in the alternative positions to those shown and with switches S_1 and S_2 closed.

The response of the system, the variation of x, $\dot{x}$ or $\ddot{x}$ with time, may then be viewed on an oscillosope or chart recorder. The effects of many different initial conditions may be readily investigated and changes in system parameters M and C may be incorporated by simply changing a resistor or the setting of a potential divider.

2.9 Experiments

(a) Use of the Simple Pendulum

Often used to measure the acceleration due to gravity, the simple pendulum may be simply constructed with a length L of light string, the top end of which is rigidly clamped and the bottom end of which is attached to a massy bob. Some 20 periods are timed and the period T calculated. The length L is then changed and the period again determined. A graph of L versus T^2 is a straight line with slope $g/4\pi^2$ from which g is found.

(b) Vertical Spring Oscillator

A hanging helical spring of constant C with a mass M attached to the bottom constitutes a harmonic oscillator with period $T = 2\pi\sqrt{(M/C)}$, if the mass of the spring can be neglected relative to M. Since by measuring the time for a large number of oscillations, T can be determined accurately, this experiment can be used to measure any one of the two quantities M or C, if the other one is known. For instance if M is known and varied, a graph of T^2 versus M yields a straight line of slope $4\pi^2/C$ from which C may be calculated.

(c) The Torsion Pendulum

A torsion pendulum consists of either a metal bar suspended at its mid-point or a cylindrical metal bob rigidly connected to a long length of wire. The upper end of the wire is rigidly fixed. The bar or bob executes rotational harmonic motion if deflected a slight amount. The period T of this harmonic motion is $T = 2\pi\sqrt{(I_b L_w / n I_w)}$ where I_b is the mass moment of inertia of the bar or bob about the axis of rotation, I_w is the area moment of inertia of the wire about the axis, n is the shear modulus of the wire material and L_w is the length of the wire. If the dimensions of the wire are carefully measured, I_w may be calculated thus, $\frac{1}{2}\pi R^4$. Then since T may be measured accurately, if n is known the moment of inertia of the bar or bob may be calculated. Alternatively if I_b is known, this experiment may be used to find the shear modulus n, of the wire material.

(d) The LC Electrical Circuit

An LC circuit such as that of figure 2.11 can be readily assembled incorporating a switch on one of the lines. Before closing the switch, various levels of D.C. voltage may be placed across the capacitor, thus charging it. Then using an oscilloscope the voltage across the capacitor may

FREE HARMONIC OSCILLATOR

be monitored after the switch is closed. The period of the oscillations may be measured. Then the value of the capacitance C or inductance L may be calculated if the other is known. For instance if C is varied a graph of T^2 versus C yields a straight line of slope $4\pi^2 L$ from which L may be calculated.

(e) Analogue Computer Circuit

Program the analogue computer as shown in figure 2.12. The response of the system to a variety of initial conditions may be monitored with an oscilloscope. Different values of M and C may be utilised by simple setting of the potentiometers.

Problems

2.1 In a system such as that of figure 2.1, the mass is 40 kg and the spring constant is 160 N/m. Calculate the frequency and the period.

2.2 If in the above system, at $t = 0$, $x = 0.2$ m and $\dot{x} = 10$ m/s, write down the expression for the displacement of the mass and calculate the total energy in the system.

2.3 A mass of 5 kg hangs on the bottom end of a vertical spring the top end of which is fixed. The period of the free oscillation is 0.5 s. Find the value of the spring constant. By how much will the spring shorten when the mass is removed? By how much should the mass be changed to produce a period of 10 s?

2.4 A mass of 100 kg hanging on a spring stretches it by 5 cm. Find the spring constant and the frequency of free oscillation of the system. If at $t = 0$, $x = 7$ cm and $\dot{x} = 10$ m/s, write down the equation for the subsequent displacement of the system.

2.5 A simple harmonic vibrator such as that of figure 2.1 has a mass of 0.2 kg and a spring of constant of 5 kN/m. The degree of freedom is horizontal. When the mass is 4 cm to the right of its equilibrium position, it has a velocity of $+2$ m/s. Write down the equation describing the displacement of the mass.

2.6 If the system at rest in problem 2.5 is given, by means of an impulse at $t = 0$, 1 J of energy, all in the kinetic form, write down the expression for the displacement of the mass. What is the expression for the force of the spring on the upright wall?

2.7 A disc of radius 1 m and mass 5 kg is mounted at its centre on the end of a circular rod of length 1 m. The radius of the rod is 1 cm and the period of torsional vibrations is 1 s. Find the shear modulus of the material of the rod.

2.8 The balance wheel of a watch has a moment of inertia of 10^{-5} kg m^2 and a period of 0.67 s. What is the torsional spring constant of the spring?

2.9 Two masses M_1 and M_2 are connected by a spring of constant C. Draw an equivalent electrical circuit for this sytem and show that the period of the oscillation of either mass is given by

$$T = 2\pi\sqrt{[C(M_1 + M_2)/M_1 M_2]}.$$

2.10 A hollow tube of diameter 30 cm is closed at each end and weighted at one end so that when placed in water it floats upright. The total mass is 200 kg. If the tube is then raised slightly and released, it oscillates. What is the period of the oscillation? Take the density of water as 10^3 kg/m^3.

2.11 If a boat is observed to have a period of free vertical oscillations of 1 s and a horizontal cross-sectional area of keel of 10 m^2, what is its mass?

2.12 A locomotive engine of mass 50 Mg is coupled to a wagon of mass 20 Mg and observed to have a frequency of free horizontal vibration of 0.2 Hz. What is the spring constant of the coupling? (See problem 2.9).

2.13 In the first complete cycle of the vibration described by $x = 5\cos\pi t + 10\sin\pi t$, what are the maximum displacement and the maximum velocity and when is the velocity a maximum?

2.14 A spring of constant C_1 hangs vertically downwards. Its lower end is attached to a spring of constant C_2 whose lower end is attached to a mass M. Draw the equivalent electrical circuit for this system and show that the period of free vibrations is

$$T = 2\pi\sqrt{[C_1 C_2/M(C_1 + C_2)]}.$$

2.15 A steel beam, three-point loaded, is 2 mm thick, 2 cm wide and 2 m long. Young's modulus for steel is 0.2×10^{12} N/m^2. A spring of constant 10 N/m hangs from the mid-point of the beam and a mass of 40 kg is attached to the lower end of this spring. What is the period of the oscillation of the mass?

2.16 A train wagon of mass 15 Mg has a frequency of free vertical vibrations on its suspension of 0.3 Hz. What is the effective spring constant of the suspension? If the suspension in fact consists of four identical springs over the four wheels what is the spring constant of each spring?

Further Reading

Bishop, R.E.D., *Vibration* (Cambridge University Press, 1965)
Braddick, H.J.J., *Vibrations, Waves and Diffraction* (McGraw-Hill, London, 1965)
Clayton, G.B., *Operational Amplifiers* (Butterworths, London, 1974)
Church, A.H., *Mechanical Vibrations,* 2nd edition (Wiley, New York, 1963)
Daish, C.B. and Fender, D.H., *Experimental Physics,* 2nd edition (Hodder & Stoughton, London, 1976)
Firth, I.M., Grant, D.F. and Wray, E.M., *Waves and Vibrations* (Penguin, London, 1973)
French, A.P., *Vibrations and Waves* (Nelson, London, 1971)
Graeme, J.G., *Applications of Operational Amplifiers* (McGraw-Hill, New York, 1974)
Haberman, C.M., *Vibration Analysis* (Merrill, Columbus, Ohio, 1968)
Kinsler, L.W. and Frey, A.R., *Fundamentals of Acoustics,* 2nd edition (Wiley, New York, 1962)
Kittel, C., Knight, W.D. and Ruderman, M.A., *Mechanics: Berkeley Physics Course,* vol 1, 2nd edition (McGraw-Hill, New York, 1973)
Morse, P.M., *Vibration and Sound,* 2nd edition (McGraw-Hill, New York, 1948)
Sears, F.W. and Zemansky, M.W., *University Physics,* 4th edition (Addison-Wesley, Reading, MA, 1972)
Seto, W.W., *Theory and Problems of Mechanical Vibrations* (Schaum, New York, 1964)
Steidel, R.F., Jr., *An Introduction to Mechanical Vibrations,* 2nd edition (Wiley, New York, 1979)
Stout, D.F. and Kaufman, M., *Handbook of Operational Amplifier Circuit Design* (McGraw-Hill, New York, 1976)
Timoshenko, S., Young, D.H. and Weaver, W. Jr., *Vibration Problems in Engineering,* 4th edition (Wiley, New York, 1974)
Tyler, F., *A Laboratory Manual of Physics,* 5th edition, SI version (Arnold, London, 1977)
Vernon, J.B., *Linear Vibration Theory,* (Wiley, New York, 1967)
Wart, J.V., Huelsman, L.P. and Korn, G.P., *Introduction to Operational Amplifier Theory and Applications* (McGraw-Hill, New York, 1975)

3 Damped Free Harmonic Oscillator

3.1 Introduction and Objectives

The aim in this chapter is to describe the free oscillatory behaviour of a simple vibratory system in which some energy loss mechanism is present.

After reading this chapter the student should be able to

(a) define and describe the various types of friction;
(b) derive the effects of dynamic friction on the motion and kinetic energy of an object;
(c) describe viscosity;
(d) outline the construction and operation of a discrete damper;
(e) derive the equation of free motion of a damped harmonic oscillator;
(f) define and describe critical damping, under-damping and over-damping;
(g) discuss the energy in a damped vibratory system;
(h) define the logarithmic decrement and the quality factor in such a system;
(i) design electrical analogue circuits for the damped harmonic oscillator.

3.2 Friction and Damping

The phenomenon of solid-body friction may be conveniently divided into three parts (a) static friction (b) dynamic friction and (c) transformation of mechanical energy into heat. Fluid friction is usually considered under the heading of 'viscosity' but in this instance heat generation is also involved.

Static or limiting friction may be described in reference to figure 3.1 which represents an object of mass M on a rough horizontal table. An external horizontal force F is applied to the mass and is gradually increased from zero. In practice it is found that for low values of F the object is not accelerated, in apparent contradiction of Newton's second law (equation 1.4). A force equal and opposite to F must be acting on the mass and the only place where this second force can physically act is at the bottom of the

DAMPED FREE HARMONIC OSCILLATOR

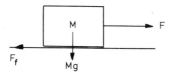

Figure 3.1 A mass M resting on a horizontal table which is subjected to an external horizontal force F experiences a frictional force F_f in a direction opposed to the force F.

body between the table and the body. This force F_f is called the static frictional force. As F is increased, as shown in figure 3.2, a value of F is eventually reached such that for any greater values of F the body experiences a linearly increasing acceleration. This behaviour is interpreted as the frictional force F_f attaining a maximum or limiting value such that when $F > F_f(\max)$ the net accelerating force is $[F - F_f(\max)]$.

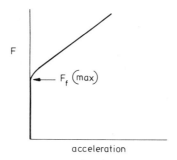

Figure 3.2 The relationship between external force and acceleration in the presence of friction. After the external force F exceeds the maximum frictional force $F_f(\max)$, acceleration increases linearly.

Such frictional force may arise from a combination of mechanical interlocking (possibly at microscopic level) between the horizontal table surface and the bottom surface of the object and chemical bonds (mainly secondary Van der Waals' or hydrogen bonds but possibly primary bonds as well) between the materials of the two surfaces in contact.

The limiting frictional force $F_f(\max)$ is found to be directly proportional to the force that the body exerts on the support surface. In the situation shown in figure 3.1, this may be written

$$F_f(\max) = \mu Mg \tag{3.1}$$

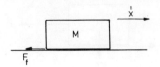

Figure 3.3 In dynamic friction, when a mass has velocity $\dot{x}$, a frictional force F_f acts to oppose the motion and is proportional to the velocity.

where μ is the coefficient of static friction and is a constant for a given object and table. μ has no dimensions.

Dynamic friction refers to the action of the fundamental frictional mechanisms when relative velocity exists between the object and the table. Consider the case shown in figure 3.3 where the body of mass M has a velocity $\dot{x}$ and is experiencing a retarding frictional force F_f. It is found experimentally that the magnitude of this retarding or braking force is proportional to the relative velocity thus

$$F_f = -b\dot{x} \tag{3.2}$$

where b is the dynamic frictional coefficient commonly called the damping coefficient. The units of b are newton seconds/metre (N s/m) or kilograms/second (kg/s).

This unbalanced force tends to decelerate the mass and so the force equation for the mass M may be written

$$-b\dot{x} = -M\ddot{x} \tag{3.3}$$

or

$$\ddot{x} - \frac{b}{M}\dot{x} = 0 \ . \tag{3.4}$$

This equation is also written

$$\ddot{x} - \frac{1}{\tau}\dot{x} = 0 \tag{3.5}$$

where

$$\tau = \frac{M}{b} \ . \tag{3.6}$$

Equation 3.5 describes an exponentially decaying motion and its solution can be shown to be

$$\dot{x} = \dot{x}(0) \, e^{-t/\tau} \tag{3.7}$$

DAMPED FREE HARMONIC OSCILLATOR

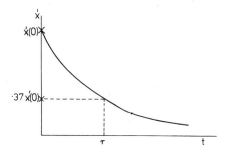

Figure 3.4 Exponential decay of velocity with time. In each characteristic time or time constant τ, the velocity falls by a factor of 0.37.

This solution is displayed in figure 3.4. Note that τ can now be seen to have the customary meaning of a relaxation time or a time constant for the motion; τ is measured in seconds (s). In τ s the velocity $\dot{x}$ declines to a fixed fraction e^{-1} ($=0.37$) of the value at the start of that interval. The complete solution given in equation 3.7 has one element τ which depends on the system parameters M and b and one element $\dot{x}(0)$ which is the initial value of $\dot{x}$, an initial condition,

The aspect of dynamic friction whereby mechanical energy is transformed into heat may be seen by considering the kinetic energy E_k of the mass.

$$\begin{aligned} E_k &= \tfrac{1}{2} M \dot{x}^2 \\ &= \tfrac{1}{2} M [\dot{x}(0)]^2 \, e^{-2t/\tau} \\ &= E_k(0) \, e^{-t/(\tau/2)} \,. \end{aligned} \tag{3.8}$$

The kinetic energy decays exponentially with a relaxation time one half of that which pertains to the decay of the velocity. As shown in figure 3.5, the kinetic energy declines twice as rapidly as the velocity.

There is no other mechanical energy storage in this system and so the mechanical energy is not conserved. The lost mechanical energy is changed into heat. Consideration of equation 3.8 shows that the rate of heat generation is greater if the relaxation time is smaller, that is, if the damping coefficient b is large or if the mass M is small.

Within a moving fluid (liquid or gas) internal friction or viscosity may be explained with reference to figure 3.6 which represents a fluid flowing over a horizontal surface. There is a velocity gradient up the height of the fluid dx/dh which will be assumed constant. Consider the two laminae of flow L_1 and L_2 with the upper one flowing at a slightly higher velocity than the

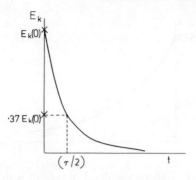

Figure 3.5 Exponential decay of kinetic energy E_k with time. The time constant is $\tau/2$ where τ is the time constant for the decay of velocity.

lower. The lower lamina then exerts a retarding, shearing stress T_{sh} on the upper lamina and it is found that this stress is directly proportional to the velocity gradient thus

$$T_{sh} = \eta \frac{dx}{dh} \tag{3.9}$$

where η is the coefficient of viscosity, a constant for the fluid. The units of η are kilogram/second square metre (kg/s m²). Viscosity results in the retardation of any relative motion within the fluid. The mechanical energy in the fluid is thereby transformed into heat.

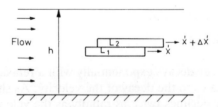

Figure 3.6 Flow of a liquid of height h over a fixed horizontal surface. Each horizontal lamina L_2 has a slightly higher velocity than the next lower lamina L_1.

The flow of a viscous liquid or gas along a tube of radius R m and length L m exerts a shear force on the internal wall of the tube. Conversely the wall exerts an equal retarding force on the liquid flow which, for the flow to persist, must be overcome by a net forward force on the liquid. A pressure difference ΔP must therefore exist between the two ends of the tube in order to maintain a steady flow rate. If $\dot{Q}$ m³/s is the steady flow rate, then Poiseuille's law describes its relation to the pressure drop thus

$$\dot{Q} = \frac{\pi R^4 \Delta P}{8 \eta L}. \tag{3.10}$$

But the flow rate is related to the average liquid velocity $\dot{x}_A$ by

$$\dot{Q} = \pi R^2 \dot{x}_A. \tag{3.11}$$

Also the net force F_L on the liquid within the tube is

$$F_L = \pi R^2 \Delta P. \tag{3.12}$$

Therefore Poiseuille's equation may be rewritten in the form

$$F_L = (8 \pi \eta L) \dot{x}_A. \tag{3.13}$$

The force exerted by the wall of the tube F_W on the liquid must be equal to, but of opposite sign to, this force thus

$$F_W = -(8 \pi \eta L) \dot{x}_A. \tag{3.14}$$

Note that this equation is identical in form with equation 3.2 characterising frictional damping between two solid bodies.

Solid-body friction between an object and the supporting table as in figure 3.3 results in dynamic damping. In such a case the value of the damping coefficient b depends on the roughnesses of the two surfaces, the chemical affinity between the materials of the surfaces, the presence of lubricants, etc. Friction between the moving object and the surrounding air contributes to the damping. The value of b may not be readily set to a desired value. Indeed, with gradual wear b will vary as time goes by. Such lack of predictable behaviour can be undesirable in a mechanical system.

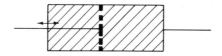

Figure 3.7 A viscous damper in which a perforated piston can be moved back and forth through a fluid.

Controlled values of damping coefficient b may be achieved with an oil or air dashpot as shown in figure 3.7. This consists of a movable perforated piston inside a cylindrical container of oil. Movement of the piston through the oil results in oil flow through the perforations and this in turn produces a viscous retarding force proportional to the piston velocity.

For example, let the thickness of the piston be L m and let there be n perforations of radius R m. Let η be the coefficient of viscosity of the oil. If the piston is moving at a steady velocity $\dot{x}$, the oil may be viewed as moving at an average velocity $-\tfrac{1}{2}\dot{x}$ through the tubes. The force exerted by the flowing oil on the piston, via the walls of the n perforation tubes, is, from equation 3.13,

$$F = -(4\pi n \eta L)\dot{x}$$
$$= -b\dot{x} . \qquad (3.15)$$

Choice of oil viscosity coefficient, number of piston perforations and piston thickness allows discrete dampers with any desired value of damping coefficient b to be designed.

Example 3.1

An object of mass 8 kg is projected horizontally along a rough surface with an initial kinetic energy of 100 J. After 10 s the velocity of the mass is 1 m/s. Find the damping coefficient between the object and the surface. What will the velocity be after a further 10 s?

Solution

The initial kinetic energy, $E_k(0) = \tfrac{1}{2}M[\dot{x}(0)]^2 = 100 = 4[\dot{x}(0)]^2$. So $\dot{x}(0) = 5$ m/s.

From equation 3.7

$$\dot{x} = \dot{x}(0)\, e^{-t/\tau}$$

where $\tau = M/b = 8/b$. When $t = 10$ s, $\dot{x} = 1$ m/s and so $1 = 5\, e^{-10/\tau}$.

$$e^{-10/\tau} = 0.2 \text{ and } \frac{10}{\tau} = 1.6 .$$

$$\frac{1}{\tau} = 0.16 = \frac{b}{8} \text{ and } b = 1.28 \text{ N s/m} .$$

When $t = 20$ s

$$\dot{x} = 5\, e^{-3.2} = 0.20 \text{ m/s} .$$

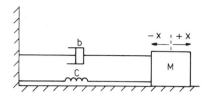

Figure 3.8 A simple harmonic oscillator with damping. The discrete element damper has a damping coefficient b.

3.3 Damped Harmonic Oscillator

Consider the simple harmonic oscillator modified by the addition of a viscous damping element as shown in figure 3.8. The conventional representation for a damping element — a piston moving in a dashpot — is used. The damping force $-b\dot{x}$ acts on the mass together with the spring force $-Cx$. The free-body diagram for the object of mass M is shown in figure 3.9 and allows the force equation for the mass to be written as

$$-b\dot{x} - Cx = M\ddot{x} . \tag{3.16}$$

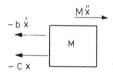

Figure 3.9 The free-body diagram for the mass in the damped harmonic oscillator. The forces due to the spring and the viscous damper result in an accelerating force.

This may be simplified to

$$\ddot{x} + \frac{1}{\tau}\dot{x} + \omega_0^2 x = 0 \tag{3.17}$$

where τ and ω_0 have been defined previously thus

$$\tau = \frac{M}{b} \tag{3.6}$$

and

$$\omega_0^2 = \frac{C}{M} \,. \tag{2.5}$$

In analogy with equation 2.10, the following may be shown to be a solution to equation 3.17

$$x = Ae^{-\alpha t}\cos(\omega_f t + \gamma) \tag{3.18}$$

where

$$\alpha = \frac{1}{2\tau} = \frac{b}{2M} \tag{3.19}$$

and

$$\omega_f = \sqrt{\left(\omega_0^2 - \alpha^2\right)}$$
$$= \sqrt{\left(\frac{C}{M} - \frac{b^2}{2M^2}\right)} \,. \tag{3.20}$$

A is the initial amplitude and γ is a phase angle. Both of these parameters depend on the initial conditions only. The coefficient of decay of the amplitude α and f_f (= $\omega_f/(2\pi)$), the frequency of the damped free vibrations of the system, are both functions of the system parameters only.

3.4 Three Different Types of Behaviour

Consideration of equation 3.20 allows three different cases.

(a) Underdamping

If

$$\alpha < \omega_0 \tag{3.21}$$

or

$$b < 2\sqrt{(CM)} \tag{3.22}$$

then

$$\omega_f < \omega_0 \,. \tag{3.23}$$

A damped sinusoidal motion as shown in figure 3.10 then occurs. The amplitude of the sinusoid decays exponentially with a time constant $1/\alpha$ or

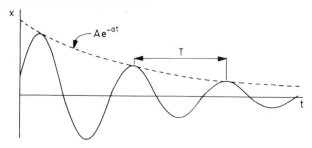

Figure 3.10 The displacement of the mass in the damped oscillator as a function of time for the case of underdamping. The amplitude of the free or natural vibration decays exponentially with a decay constant α or time constant $1/2\alpha$.

2τ. The greater the amount of damping, the more rapid is the decline of the amplitude and the lower the frequency of the free vibration.

In the $\dot{x}/x$ phase plane plot for this system, the locus describing the free vibratory behaviour is the reducing spiral shown in figure 3.11. In this diagram P is the initial state of the system and the spiral is traced out by a radial line rotating clockwise at ω_f rad/s.

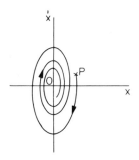

Figure 3.11 A mass velocity versus mass displacement plot for underdamped free vibration. The locus starts at the initial conditions and traces a reducing spiral in the clockwise direction.

If the damping is very slight and

$$\alpha \ll \omega_0 \text{ or } b \ll 2\sqrt{(CM)}$$

then

$$\omega_f = \omega_0$$

and equation 3.18, describing the displacement of the mass as a function of time, becomes

$$x = A\, e^{-\alpha t} \cos(\omega_0 t + \gamma) \ . \tag{3.24}$$

In this case the frequency is essentially the undamped natural frequency and the amplitude declines with a time constant 2τ.

(b) Critical Damping

If

$$\alpha^2 = \omega_0^2 \tag{3.25}$$

or

$$b = 2\sqrt{(CM)}$$
$$= b_c \tag{3.26}$$

where this value of damping coefficient b_c produces critical damping, then

$$\omega_f = 0 \ . \tag{3.27}$$

Such a system which does not oscillate is said to be critically damped. In this case the solution of equation 3.18 to the equation of motion of the system becomes

$$x = A\, e^{-\alpha_c t} \cos \gamma$$
$$= A'\, e^{-\alpha_c t} \ . \tag{3.28}$$

Here A' is a constant amplitude term set by the initial conditions and

$$\alpha_c = \frac{b_c}{2M}$$
$$= \sqrt{\left(\frac{C}{M}\right)}$$
$$= \omega_0 \ . \tag{3.29}$$

In this case the following may also be shown to be a solution to the basic equation of motion

$$x = Bt\, e^{-\alpha_c t} \tag{3.30}$$

DAMPED FREE HARMONIC OSCILLATOR

where B is also a constant amplitude term determined by the initial conditions. The sum of the two solutions in equations 3.28 and 3.30 is also a solution and therefore

$$x = (A' + Bt)\,e^{-\alpha_c t}. \tag{3.31}$$

This equation describes the displacement x for the critically damped oscillator freely moving. The variation of the displacement with time is shown in figure 3.12. There is no oscillation, only a less gradual decline than an exponential decay of decay constant α_c.

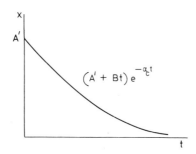

Figure 3.12 The decay of the mass displacement in a critically damped oscillator, freely moving.

(c) Overdamping

If

$$\alpha^2 > \omega_0^2 \tag{3.32}$$

or

$$b > 2\sqrt{(CM)} \tag{3.33}$$

the system is said to be overdamped. Then

$$\omega_f = \sqrt{\omega_0^2 - \alpha^2}$$

$$= j\sqrt{\alpha^2 - \omega_0^2} \tag{3.34}$$

$$= j\psi. \tag{3.35}$$

In this case the solution to the equation of motion 3.17 may be shown by substitution to take the form

$$x = (A\, e^{\alpha t} + B\, e^{-\alpha t})\, e^{-\psi t}. \tag{3.36}$$

This displacement is plotted as a function of time in figure 3.13, and it is a monotonically declining function with a rate of decay slower even than that for critical damping. There is no oscillation. The constants A and B are determined by the initial conditions while α and ψ are set by the system parameters.

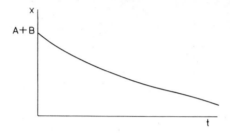

Figure 3.13 Displacement of the mass as a function of time in the freely moving overdamped harmonic oscillator.

The damping coefficient is often expressed as a ratio of the value for critical damping. Thus the damping ratio q is defined

$$q = \frac{b}{b_c}$$

$$= \frac{\alpha}{\omega_0} \tag{3.37}$$

$$= \frac{b}{2\sqrt{(CM)}}. \tag{3.38}$$

In the case of underdamping $q < 1$, for critical damping $q = 1$ and for overdamping $q > 1$.

Example 3.2

In a mechanical system such as that of figure 3.8 the mass is 4 kg and the spring constant is 25 N/m. Three values of damping coefficient, $0.1\, b_c$, b_c and $10\, b_c$ are to be evaluated. In each case the mass is displaced by +1 m and released from rest. Find how long it takes in each case for the maximum amplitude of the motion to become less than 0.1 m.

DAMPED FREE HARMONIC OSCILLATOR

Solution

In this system

$$\omega_0 = \sqrt{\left(\frac{C}{M}\right)} = 2.5 \text{ rad/s} .$$

For critical damping, from equation 3.26 $b_c = 2\sqrt{(CM)} = 20$ kg/s.

(a) Underdamping

Here $b = 2$ kg/s and from equation 3.19

$$\alpha = \frac{b}{2M} = 0.25 \text{ 1/s} .$$

From equation 3.18, $x = A\,e^{-\alpha t}\cos(\omega_f t + \gamma)$ and $\dot{x} = -\alpha A\,e^{-\alpha t}\cos(\omega_f t + \gamma) - \omega_f A\,e^{-\alpha t}\sin(\omega_f t + \gamma)$ where

$$\omega_f = \sqrt{(\omega_0^2 - \alpha^2)} = 2.49 \text{ rad/s} .$$

At $t = 0$, $x(0) = 1 = A\cos\gamma$ and $\dot{x}(0) = 0 = -\alpha A\cos\gamma - \omega_f A\sin\gamma$. So

$$\tan\gamma = -\frac{\alpha}{\omega_f} = -0.1 \text{ and } \gamma = 0.1; \cos\gamma = 0.99$$

For the amplitude of x to be 0.1 m, $A\,e^{-\alpha t} = 0.1$ or $1.01\,e^{-0.25t} = 0.1$. Then $t = 9.3$ s.

(b) Critical Damping

Now $b = b_c = 20$ kg/s. From equation 3.31

$$x = (A + Bt)\,e^{-\alpha_c t} \text{ and } \dot{x} = (-\alpha_c A - \alpha_c B t + B)\,e^{-\alpha_c t}$$

where $\alpha_c = \omega_0 = 2.5$ 1/s from equation 3.25 .
At $t = 0$, $x(0) = A = 1$ m and $\dot{x}(0) = 0 = -\alpha_c A + B$. So

$$B = \alpha_c A = 2.5 \text{ m/s} .$$

Now for $x = 0.1$ m

$$(1 + 2.5t)\,e^{-2.5t} = 0.1 .$$

By trial and error, $t = 1.54$ s.

(c) Overdamping

Here $b = 200$ kg/s and $\alpha = 200/8 = 25$ 1/s. From equation 3.35

$$\psi = \sqrt{(\alpha^2 - \omega_0^2)} = 24 \ .$$

From equation 3.36

$$x = A\, e^{(\alpha-\psi)t} + B\, e^{-(\alpha+\psi)t}$$

and

$$\dot{x} = (\alpha - \psi)A\, e^{(\alpha-\psi)t} - (\alpha + \psi)B\, e^{-(\alpha+\psi)t} \ .$$

But at $t = 0$

$$x(0) = 1 \text{ m} = A + B \text{ and}$$
$$\dot{x}(0) = (\alpha - \psi)A - (\alpha + \psi)B$$
$$= -0.13A - 49.87B = 0$$

So

$$B = -0.0026 A$$

and

$$A = 1.003 \text{ m and } B = -0.003 \text{ m} \ .$$

For

$$x = 0.1 \text{ m}$$

$$0.1 = 1.003\, e^{-0.13t} - 0.003\, e^{-49.87t}$$
$$= 1.003\, e^{-0.13t}$$

Therefore

$$t = 0.04 \text{ s}$$

3.5 Energy in the Damped System

In the damped oscillatory system the two forms of energy present, the potential E_p and kinetic E_k, may be written as before

DAMPED FREE HARMONIC OSCILLATOR

$$E_p = \tfrac{1}{2}Cx^2$$

and

$$E_k = \tfrac{1}{2}M\dot{x}^2 .$$

The displacement x is given by equations 3.18, 3.31 and 3.36 depending on whether the system is underdamped, critically damped or overdamped.

Consider the case of underdamping in which vibrations occur. Here

$$x = A\, e^{-\alpha t} \cos(\omega_f t + \gamma) . \tag{3.18}$$

After differentiation one obtains the velocity $\dot{x}$

$$\dot{x} = -A\, e^{-\alpha t}[\alpha \cos(\omega_f t + \gamma) + \omega_f \sin(\omega_f t + \gamma)]$$

$$= -\omega_0 A\, e^{-\alpha t} \cos\left[\omega_f t + \gamma - \tan^{-1}\left(\frac{\omega_f}{\alpha}\right)\right] . \tag{3.39}$$

Therefore the potential energy E_p in the system at any instant is given by

$$E_p = \tfrac{1}{2}CA^2\, e^{-2\alpha t} \cos^2(\omega_f t + \gamma) . \tag{3.40}$$

The kinetic energy in the system E_k is given by

$$E_k = \tfrac{1}{2}M\omega_0^2 A^2\, e^{-2\alpha t} \cos^2\left[\omega_f t + \gamma - \tan^{-1}\left(\frac{\omega_f}{\alpha}\right)\right]$$

$$= \tfrac{1}{2}CA^2\, e^{-2\alpha t} \cos^2\left[\omega_f t + \gamma - \tan^{-1}\left(\frac{\omega_f}{\alpha}\right)\right] . \tag{3.41}$$

The instantaneous total mechanical energy E_t in the system is the sum of these two forms

$$E_t = E_p + E_k$$

$$= \tfrac{1}{2}CA^2\, e^{-2\alpha t}\left(1 + \frac{\alpha}{\omega_0} \cos\left[2\omega_f t + 2\gamma - \tan^{-1}\left(\frac{\omega_f}{\alpha}\right)\right]\right) \tag{3.42}$$

Figure 3.14 illustrates how this total energy varies with time. It declines steadily but with a constantly undulating slope. The frequency of these undulations is twice that of the free damped oscillations of the system.

The rate at which the total mechanical energy falls or is lost from the system is the rate of heat production by the damping mechanism. If H is the heat produced, then

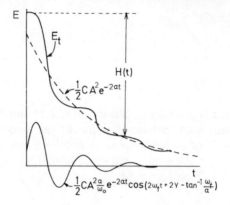

Figure 3.14 The decline of the total energy in the system E_t as a function of time in the underdamped oscillator, freely vibrating. The heat produced up to each instant $H(t)$ is also shown to be equal to the reduction in total mechanical energy up to that instant.

$$\dot{H} = -\dot{E}_t \qquad (3.43)$$
$$= -(\dot{E}_p + \dot{E}_k)$$
$$= -[\tfrac{1}{2}C(2x\dot{x}) + \tfrac{1}{2}M(2\dot{x}\ddot{x})]$$
$$= -\dot{x}(Cx + M\ddot{x})$$
$$= -\dot{x}(-b\dot{x})$$
$$= b\dot{x}^2 . \qquad (3.44)$$

Thus the rate of heat production may be found by directly differentiating the expression for the total energy E_t in equation 3.42 or by substituting the expression for the velocity of the mass $\dot{x}$ from equation 3.39 into equation 3.44 thus

$$\dot{H} = b\omega_0^2 A^2 \, e^{-2\alpha t} \cos^2\left[\omega_f t + \gamma - \tan^{-1}\left(\frac{\omega_f}{\alpha}\right)\right] . \qquad (3.45)$$

Both approaches can be shown to yield the same result.

Note that the total amount of heat produced in the damper may be found from equation 3.43 by integrating both sides of that equation, thus

$$H = K - E_t \qquad (3.46)$$

where K is a constant. Since $H = 0$ at $t = 0$, that is, no heat is produced before the motion starts, then

DAMPED FREE HARMONIC OSCILLATOR

$$K = E_t(0)$$

$$= \tfrac{1}{2}CA^2\left(1 + \frac{\alpha}{\omega_0} \cos\left[2\gamma - \tan^{-1}\left(\frac{\omega_f}{\alpha}\right)\right]\right) \tag{3.47}$$

The heat produced up to any instant may be found from figure 3.14. It is the amount of energy lost from the start up to t.

Example 3.3

In the system of example 3.2 what is the initial energy and the total energy when the maximum displacement is 0.1 m for each of the three degrees of damping?

Solution

(a) Underdamping

At $t = 0$, all the energy is potential, so

$$E_t(0) = E_p(0) = \tfrac{1}{2}C[x(0)]^2 = 12.5 \text{ J} .$$

At $t = 9.3$ s, equation 3.42 may be used to calculate E_t. In that equation

$\omega_0^2 = 6.25$ and $\omega_0 = 2.5$ rad/s
$C = 25$ N/m
$A = 1.01$ m and $A^2 = 1.02$ m^2
$\alpha = 0.25$ 1/s and so $\alpha t = 2.33$
$\dfrac{\alpha}{\omega_0} = 0.1$
$\omega_f = 2.49$ rad/s
$\gamma = 0.1$ and $2\gamma = 0.2$ rad
$\tan^{-1}(\omega_f/\alpha) = 1.47$ rad
$\cos[2\omega_f t + 2\gamma - \tan^{-1}(\omega_f/\alpha)] = 0.49$.

Then from equation 3.42

$$E_t = 0.5 \times 25 \times 1.02 \text{ e}^{-4.66}(1.049)$$
$$= 0.13 \text{ J about 1 per cent of the original amount}$$

(b) Critical Damping

At $t = 0$, $E_t = E_p = 12.5$ J, as above. At $t = 1.54$ s, from equation 3.31

$$x = (1.0 + 2.5 \times 1.54)\, e^{-2.5 \times 1.54} = 0.1 \text{ m}$$
$$\dot{x} = (-2.5 - 2.5^2 \times 1.54 + 2.5)\, e^{-2.5 \times 1.54} = -0.204 \text{ m/s}^2 .$$

Then
$$E_p = 0.5 \times 25 \times 0.1^2 = 0.13 \text{ J}$$
$$E_k = 0.5 \times 4 \times (0.204)^2 = 0.08 \text{ J}$$
$$E_t = 0.21 \text{ J about 1.7 per cent of the original amount.}$$

(c) Overdamping

At $t = 0$, $E_t = 12.5$ J, as in the two earlier cases. At $t = 0.04$ s, $x = 0.1$ m and

$$\dot{x} = -0.13 \times 1.003\, e^{-0.13 \times 0.04} + 49.87 \times 0.003\, e^{-49.87 \times 0.04}$$
$$= -0.11 \text{ m/s} .$$

Then
$$E_p = 0.5 \times 25 \times 0.1^2 = 0.13 \text{ J}$$
$$E_k = 0.5 \times 4 \times 0.11^2 = 0.02 \text{ J}$$

So
$$E_t = 0.15 \text{ J, about 1.2 per cent of the original total energy.}$$

3.6 Functional Parameters of the Underdamped System

Often in a damped oscillatory system, some or all of the basic parameters M, b and C may not be known in value. Furthermore it may not be possible to measure these quantities accurately. In such circumstances, monitoring and measuring the performance of the system allows a number of derived or functional parameters of the system to be calculated.

One of these parameters is the logarithmic decrement δ which is the relative decline of the amplitude of the free oscillation of figure 3.10, in one period T. Thus

$$\delta = \frac{A e^{-\alpha t} - A e^{-\alpha(t+T)}}{A e^{-\alpha t}}$$
$$= 1 - e^{-\alpha T} . \tag{3.48}$$

DAMPED FREE HARMONIC OSCILLATOR

If the damping is very slight, $\alpha T \ll 1$, that is, $\alpha \ll \omega_0/(2\pi)$ or $b \ll (1/\pi)\sqrt{(CM)}$, then

$$e^{-\alpha T} = 1 - \alpha T$$

and

$$\delta = \alpha T \tag{3.49}$$

$$= \frac{2\pi\alpha}{\omega_0} \tag{3.50}$$

$$= \frac{\pi b}{\sqrt{(CM)}} \tag{3.51}$$

$$= 2\pi q \tag{3.52}$$

where q is the damping ratio defined in equation 3.37.

In a lightly damped system the logarithmic decrement δ is small. An alternative and almost reciprocal parameter is the quality factor Q of the system. A number of formulations for Q are useful thus

$$Q = \frac{\pi}{\delta} \tag{3.53}$$

$$= \frac{\omega_0}{2\alpha} \tag{3.54}$$

$$= \frac{\sqrt{(CM)}}{b} \tag{3.55}$$

$$= \frac{1}{2q} \,. \tag{3.56}$$

A heavily damped system has a low Q value while a high Q value is indicative of a lightly damped system.

From equation 3.42, in a lightly damped system the fractional reduction in total energy E_t, in each period T is given by

$$1 - e^{-2\alpha T} = 1 - (1 - 2\alpha T)$$
$$= 2\alpha T$$
$$= \frac{2\pi}{Q} \tag{3.57}$$

In one period T the energy declines by a factor $2\pi/Q$. This is another often used definition of the Q value or Q number of a system. Some of the important applications of this parameter will be discussed in chapter 4.

Example 3.4

In a freely vibrating mechanical system the ratio of two successive amplitudes of the motion is 10/9 and these amplitudes are separated by 1 s. The mass involved is 100 kg. Find the logarithmic decrement, the damping ratio, the Q value and the damping coefficient.

Solution

The logarithmic decrement δ is given by

$$\delta = \frac{10 - 9}{10} = 0.1 \ .$$

The system is underdamped and so assume $\delta = \alpha T$ where $T = 1$ s. So using this information and equation 3.19

$$\alpha = 0.1 = \frac{b}{2M} \text{ so that } b = 20 \text{ kg/s} \ .$$

From equation 3.52, the damping ratio q is found

$$q = \frac{\delta}{2\pi} = 0.016 \ .$$

From equation 3.53, the quality factor Q may be calculated

$$Q = \frac{\pi}{\delta} = 31.4 \ .$$

3.7 Electrical Analogue Circuits

The introduction of a discrete lossy element, a resistance R, into the circuit of figure 2.11, as shown in figure 3.15, produces a circuit with the basic voltage equation

$$L\ddot{q} + R\dot{q} + \frac{q}{C} = 0 \qquad (3.58)$$

where q is here electrical charge. Dividing each term by L

$$\ddot{q} + \frac{R}{L}\dot{q} + \frac{q}{LC} = 0 \qquad (3.59)$$

or

$$\ddot{q} + \frac{1}{\tau}\dot{q} + \omega_0^2 q = 0 \ . \qquad (3.60)$$

DAMPED FREE HARMONIC OSCILLATOR

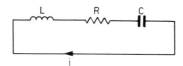

Figure 3.15 A short-circuited series resistor/inductor/capacitor circuit which is an exact analogue of the damped harmonic oscillator of figure 3.8.

This is an exact analogy with equation 3.17 where in this case

$$\tau = \frac{L}{R} \tag{3.61}$$

$$\omega_0^2 = \frac{1}{LC} \tag{3.62}$$

and

$$\alpha = \frac{R}{2L}. \tag{3.63}$$

In fact the basic set of analogous quantities are listed in table 2.1. Using these analogies the solutions to equation 3.60 can be written down from equation 3.18 for underdamping, equation 3.31 for critical damping and equation 3.36 for overdamping. The various functional parameters can also be written. Thus the quality factor

$$Q = \frac{1}{R}\sqrt{\left(\frac{L}{C}\right)} \tag{3.64}$$

and the logarithmic decrement for the electrical analogue circuit is

$$\delta = \pi R \sqrt{\left(\frac{C}{L}\right)}. \tag{3.65}$$

Once more, since the electrical circuit can be assembled so readily and the various parameters altered so easily, it is recommended in general that the electrical analogue be examined in detail before a mechanical design be finalised.

A more powerful and adaptable electrical analogue can be constructed using analogue computer techniques. Thus in order to simulate the mechanical system of figure 3.8, characterised by equations 3.16 and 3.17, the analogue computer scheme of figure 3.16 may be used. This scheme is similar to that in figure 2.12 with the addition of a potential divider to produce a fraction b/M of $-\dot{x}$ and operational amplifier number 4 which is connected as an inverting amplifier of unity gain. The output of this

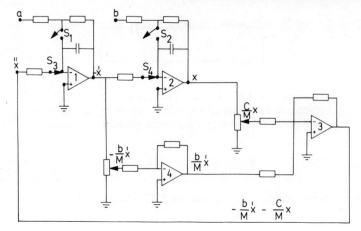

Figure 3.16 An analogue computer system which may be used to solve equation 3.17, the equation of motion of the damped harmonic vibrator in free motion. (After Clayton).

inverter $(b/M)\dot{x}$ is mixed with $(C/M)x$ to form the input to the final inverter (operational amplifier number 3). Thus the feedback line carries $-[(b/M)\dot{x} + (C/M)x]$ which equation 3.16 indicates is equal to $\ddot{x}$.

This scheme requires only that potential divider settings be varied to effect different values of C/M and/or of b/M. Different initial conditions $\dot{x}(0)$ or $\dot{q}(0)$ and $x(0)$ or $q(0)$ may be achieved by applying different voltage levels at points a and b respectively.

3.8 Experiments

(a) Measurement of Static Friction Coefficient

To measure the coefficient of static friction a flat slab that can slide on a horizontal table is used. A string is attached to the slab and passes over a pulley to a pan which can be loaded with masses. Also the mass of the slider may be increased by placing masses on its surface. For a given mass of slider the mass in the pan is gradually increased until the slider barely moves. The weight of this mass is then the maximum frictional force for the given slider mass. This slider mass is increased and a new maximum frictional force determined. A graph of maximum frictional force versus weight of slider is a straight line of slope μ, the static frictional coefficient. Different contact surfaces can be investigated in this way as can the use of lubricants between the surfaces.

(b) Study of Flow of a Viscous Liquid along a Capillary Tube

The liquid under study is made to flow under a constant head through a tube of length L and radius R. Each of these dimensions must be carefully measured, especially the radius, since it appears to the fourth power in the flow formula, Poiseuille's equation. If the head is expressed as a ΔP, the volume carried through over say 100 s can be measured in a graduated cylinder. The flow rate Q can be calculated and is given by $Q = \Delta P \pi R^4/(8\eta L)$ where η is the coefficient of viscosity of the liquid. If ΔP is varied and Q measured for each ΔP, a graph of Q versus ΔP is a straight line of slope $\pi R^4/(8\eta L)$. η may be calculated from this slope since the tube dimensions are known.

Tubes of different lengths and radii may be used as can liquids of different viscosities. Similar experiments can be carried out using gases as the flowing fluids.

(c) Damped Simple Pendulum

If the bob of the simple pendulum is made to rub against a rough vertical surface as it oscillates, the motion will be damped simple harmonic. If different grades of sandpaper are used on the rough surface, different damping coefficients can be achieved. If marks are placed on the paper at one half of the original amplitude, and the time for the oscillations to decay to this level are measured, then the decay constant α of the oscillation can be calculated. The damped frequency can be calculated from the period as can the undamped frequency. Using equation 3.20 the decay constant may also be calculated from these two frequencies. Alternatively the free undamped frequency may be calculated from the measured damped frequency and the decay constant.

(d) Damped Vertical Spring Oscillator

Variable damping can be incorporated into the vertical spring oscillator by means of a flat perforated disc mounted horizontally and rigidly attached to the bottom of the mass-bearing pan. This disc is allowed to protrude into the open surface of a viscous oil or other liquid. The damping coefficient may be varied, following equation 3.15, by changing the oil, the thickness of the disc, the number of perforations and the radii of these perforations. The damped frequency may be readily measured as can the decay constant in a manner similar to that described above. As well as allowing the damping coefficient to be varied, the mass M may be easily changed as can the spring and hence the spring constant C. Many different initial conditions may also be used.

(e) Damped Torsional Vibrator

Damping may be introduced into the torsional vibrator in a manner similar to the above except that the moving element in the viscous oil must be a vane with perforations. As the vane rotates with the bob, the flow of the liquid through the perforations results in viscous damping.

(f) Damped Simple Pendulum (Eddy Current Damping)

If the form of the mass of the simple pendulum is a metallic disc it can be made to oscillate through the poles of a magnet — either a permanent magnet or an electromagnet. As the metal disc swings through the magnetic field, eddy currents are induced in it producing braking action and mechanical energy loss. If an electromagnet is used, the damping coefficient may be varied by varying the current and so the magnetic field strength.

(g) Electrical *RLC* Circuit

The circuit of figure 3.15 may be connected with desired values of R, L and C and a switch in the circuit. Initial charge can be stored on the capacitor and the behaviour of the circuit studied after the switch is closed, using an oscilloscope. The voltage across the resistor gives a picture of the current flow. The voltage across the capacitor gives a picture of the instantaneous charge q, the analogue of x. The damped natural frequency and the decay constant may be derived from the oscilloscope display. If one of the circuit parameters is unknown and the other two known, the third may then be calculated. Many different initial conditions may be investigated.

(h) Analogue Computer Set-up

The analogue computer may be programmed as in figure 3.16, using the desired values of b, M and C. Many different initial conditions may be readily used and the performance studied with an oscilloscope. Either $\ddot{x}$, $\dot{x}$ or x can be viewed instantaneously.

Problems

3.1 A mass of 50 kg on an inclined plane of angle 30° needs a force of 10 N upwards along the plane to prevent it from slipping. What is the coefficient of static friction between the two contact surfaces? For the same two surfaces the angle of the plane is reduced until a force of 10 N down the plane is needed to cause it to slip. What is the new angle of the plane?

DAMPED FREE HARMONIC OSCILLATOR

3.2 A mass of 10 kg has an initial kinetic energy of 250 J and after 5 s of motion along the horizontal the velocity is 2 m/s. What is the damping coefficient? Write down the expression for the braking force.

3.3 A uniform ladder of length 4 m and mass 25 kg leans against a wall making an angle of 30° with the wall. The wall exerts a force on the top of the ladder perpendicular to the ladder. Find the magnitude of this force and the coefficients of friction at the wall and at the ground.

3.4 An oil of viscosity coefficient 0.1 kg/m s is flowing at a steady rate through a tube of radius 1 mm and length 1 cm. A volume of 10^{-3} m^3 passes through each minute. What force does this flow exert on the inner wall of the tube?

3.5 A dashpot is to be designed using the same oil as in the previous problem and with damping coefficient 1 kg/s. How many identical circular perforations in a piston of thickness 1 cm must be used?

3.6 In a vibratory system such as that in figure 3.8 the mass is 0.1 kg and the spring has a constant 10 N/m. Find the value of damping coefficient for critical damping. If the damping ratio is set at 0.5, and if at $t = 0$ the displacement is 0.2 m and the mass is motionless, write down the expression for the subsequent displacement of the mass.

3.7 If in the system of problem 3.6 a steady force of 5 N is applied at $t = 0$, write down the expression for the displacement.

3.8 A mass of 16 kg hangs from a spring of constant 900 N/m and there is a damping ratio of 0.1 in the system. Calculate the damping coefficient and the frequency of the damped free vibration of the mass.

3.9 In the system of problem 3.8 at $t = 0$ the mass is pulled down an additional 10 cm from its equilibrium location and given an upward velocity of 2 m/s. Find the expression for the subsequent displacement of the mass.

3.10 A mass of 10 kg hangs on a spring. When an additional mass of 1 kg is added the spring extends an additional 1 cm. The damping ratio is 0.2. The 1 kg mass is suddenly removed at $t = 0$. What is the subsequent displacement of the mass?

3.11 Plot a graph of the percentage difference between the damped and undamped natural frequencies as a function of damping ratio.

3.12 The ratio of two successive amplitudes of a damped free oscillation is 1.5. Calculate the logarithmic decrement and the damping ratio.

3.13 A door of 40 kg and 1 m wide has an automatic closer. If the torsion spring has a constant of 20 N m/rad what damping coefficient is needed for critical damping? If this door is opened to 0.5π and released from rest at $t = 0$ how long will it take for the opening to be less than 0.005π?

3.14 A mass of 100 kg is attached to a spring as in figure 3.8 and in the absence of damping demonstrates a period of vibration of 0.1 s. When damping is added, the period becomes 0.105 s. Calculate the spring constant, the damping coefficient, the damping ratio and the logarithmic decrement of the system.

3.15 If the system in problem 3.14 is equipped with an adjustable damper, draw a graph of the displacement of the mass from an initial displacement of 1 mm at rest for the damping ratio equal to (a) 0.9, (b) 1 and (c) 1.1.

3.16 Draw an electrical circuit diagram to simulate the arrangements of the system in problem 3.15.

3.17 A shock absorber is essentially a spring in parallel with a damper. A railway buffer has a spring of constant 150 kN/m as part of the shock absorber. Find the damping coefficient that will ensure that when a wagon of 20 Mg strikes the buffer the recoil will be less than 5 per cent of the maximum displacement of the absorber.

Further Reading

Anderson, R.A., *Fundamentals of Vibration* (Macmillan, New York, 1967)
Bowden, F.P. and Tabor, D., *Friction: An Introduction to Tribology* (Heinemann, London, 1973)
Church, A.H., *Mechanical Vibrations*, 2nd edition (Wiley, New York, 1963)
Daish, C.B. and Fender, D.H., *Experimental Physics*, 2nd edition (Hodder & Stoughton, London, 1976)
Dimarogonas, A.D., *Vibration Engineering* (West, St. Paul, 1976)
French, A.P., *Vibrations and Waves* (Nelson, London, 1971)
Graeme, J.G., *Applications of Operational Amplifiers* (McGraw-Hill, New York, 1974)
Haberman, C.M., *Vibration Analysis* (Merrill, Columbus, 1968)
Kinsler, L.W. and Frey, A.R., *Fundamentals of Acoustics*, 2nd edition (Wiley, New York, 1962)
Kittel, C., Knight, W.D. and Ruderman, M.A., *Mechanics: Berkeley Physics Course*, vol 1, 2nd edition (McGraw-Hill, New York, 1973)

Main, I.G., *Vibrations and Waves in Physics*, (Cambridge University Press, Cambridge, 1978)

Morse, P.M., *Vibration and Sound,* 2nd edition (McGraw-Hill, New York, 1948)

Seto, W.W., *Theory and Problems of Mechanical Vibrations* (Schaum, New York, 1964)

Steidel, R.F., Jr., *An Introduction to Mechanical Vibrations*, 2nd edition (Wiley, New York, 1979)

Stout, D.F. and Haufman, M., *Handbook of Operational Amplifier Circuit Design* (McGraw-Hill, New York, 1976)

Timoshenko, S., Young, D.H. and Weaver, D. Jr., *Vibration Problems in Engineering,* 4th edition (Wiley, New York, 1974)

Tyler, F., *A Laboratory Manual of Physics,* 5th edition, SI version (Arnold, London, 1977)

Vernon, J.B., *Linear Vibration Theory* (Wiley, New York, 1967)

Wart, J.V., Huelsman, L.P. and Korn, G.A., *Introduction to Operational Amplifier Theory and Applications* (McGraw-Hill, New York, 1975)

4 Sinusoidal Driving of the Damped Harmonic Oscillator

4.1 Introduction and Objectives

The aim of this chapter is to describe the response of the damped harmonic oscillator to an externally applied sinusoidal force.

After reading this chapter the student should be able to

(a) describe the transient and steady-state responses of the oscillator when acted on by a sinusoidal force;
(b) derive the resonant response of the amplitude of the steady-state displacement function and of the phase angle of that response;
(c) define mechanical impedance and describe the components of this parameter;
(d) outline the energy and power relationships in the driven system;
(e) derive the power resonant behaviour of the system;
(f) show the importance of the quality factor of the system in this latter resonance;
(g) outline electrical resonant circuit analogues for the driven mechanical harmonic oscillator.

4.2 Sinusoidally Forced Harmonic Oscillator

Consider the damped harmonic oscillator with one degree of freedom along the x-axis in which the body of mass M experiences an externally applied horizontal force along the x-direction as shown in figure 4.1. The free-body diagram for the mass is shown in figure 4.2 and it allows the force equation for the body to be written thus

$$f(t) - bx' - Cx = Mx'' \ . \tag{4.1}$$

After division of each term by M this equation reduces to

$$x'' + 2\alpha x' + \omega_0^2 x = g(t) \tag{4.2}$$

SINUSOIDAL DRIVING OF THE DAMPED HARMONIC OSCILLATOR

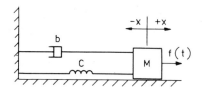

Figure 4.1 A damped harmonic oscillator with one degree of freedom along the x-axis driven by an external force $f(t)$.

where

$$g(t) = \frac{f(t)}{M}$$

$$\alpha = \frac{b}{2M} \tag{3.19}$$

and

$$\omega_0^2 = \frac{C}{M}. \tag{2.5}$$

Assume that $f(t)$ is sinusoidal thus

$$f(t) = F_0 \cos \omega t \tag{4.3}$$

such that

$$g(t) = G_0 \cos \omega t \tag{4.4}$$

where

$$G_0 = \frac{F_0}{M}. \tag{4.5}$$

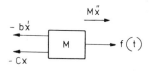

Figure 4.2 The free-body diagram for the mass in the driven damped harmonic oscillator. The resultant of the spring and damper forces and the external force accelerates the mass.

It is found that there are two parts to the solution of equation 4.2, a transient response x_t and a steady-state response x_s. The transient response is the solution to the equation when $g(t) = 0$ in which case equation 4.2 reduces to equation 3.17. The steady-state response is the response of the system to the sinusoidal $f(t)$. Both responses occur but the transient portion is short-lived so that after it has decayed appreciably only the steady-state response exists.

(a) Transient Response

This part of the response is the decaying motion described in chapter 3. For an underdamped system, of especial interest here, in which the damping ratio $q < 1$

$$x_t = A_t \, e^{-\alpha t} \cos(\omega_f t + \gamma_t) \tag{4.6}$$

where

$$\omega_f^2 = \omega_0^2 - \alpha^2 \tag{3.20}$$

and the amplitude A_t and phase angle γ_t are set by the initial conditions. For a critically damped system in which $q = 1$

$$x_t = (A_t' + B_t t) \, e^{-\alpha t} \tag{3.31}$$

where the amplitude terms A_t' and B_t are set by the initial conditions. For an overdamped system in which $q > 1$

$$x_t = (A_t \, e^{\alpha t} + B_t \, e^{-\alpha t}) \, e^{-\psi t} \tag{3.36}$$

where

$$\psi = -j\omega_f \tag{3.35}$$

and the two amplitude terms A_t and B_t are determined by the initial conditions.

(b) Steady-state Response

This response is a steady-state oscillation of the same frequency as the driving force. The following trial solutions might be used

$$x_s = A_s \cos \omega t + B_s \sin \omega t \tag{4.7}$$

SINUSOIDAL DRIVING OF THE DAMPED HARMONIC OSCILLATOR

or

$$x_s = C_s \cos(\omega t - \gamma_s) . \tag{4.8}$$

These two solutions are equivalent if

$$C_s = \sqrt{(A_s^2 + B_s^2)} \tag{4.9}$$

and

$$\gamma_s = \tan^{-1}\left(\frac{B_s}{A_s}\right) . \tag{4.10}$$

To find the appropriate values of A_s and B_s in trial solution 4.7 it must be substituted into equation 4.2. To do so the velocity $\dot{x}$ and the acceleration $\ddot{x}$ must first be known. By differentiation of equation 4.7,

$$\dot{x} = -\omega(A_s \sin \omega t - B_s \cos \omega t) \tag{4.11}$$

and

$$\ddot{x} = -\omega^2(A_s \cos \omega t + B_s \sin \omega t) . \tag{4.12}$$

Substitution of the appropriate terms in equation 4.2 and grouping the sine and cosine terms yield

$$(-\omega^2 A_s + 2\alpha\omega B_s + \omega_0^2 A_s - G_0) \cos \omega t$$
$$+ (-\omega^2 B_s - 2\alpha\omega A_s + \omega_0^2 B_s) \sin \omega t = 0 .$$

The coefficients of both $\cos \omega t$ and $\sin \omega t$ must therefore each equal zero. Thus

$$(\omega_0^2 - \omega^2)B_s - 2\alpha\omega A_s = 0 \tag{4.13}$$

and

$$(\omega_0^2 - \omega^2)A_s + 2\alpha\omega B_s = G_0 . \tag{4.14}$$

From equation 4.13

$$B_s = \frac{2\alpha\omega A_s}{\omega_0^2 - \omega^2} \tag{4.15}$$

and so equation 4.14 reduces to

$$A_s = \left[\frac{\omega_0^2 - \omega^2}{(\omega_0^2 - \omega^2)^2 + (2\alpha\omega)^2}\right] G_0 \qquad (4.16)$$

and equation 4.15 to

$$B_s = \left[\frac{2\alpha\omega}{(\omega_0^2 - \omega^2)^2 + (2\alpha\omega)^2}\right] G_0 . \qquad (4.17)$$

Thus the steady-state response has one component with amplitude A_s exactly in phase with the driving force and a second component of amplitude B_s which is $\pi/2$ rad out of phase with and lagging the driving force. Each component has the same frequency as the driving force.

Both of the amplitude terms A_s and B_s are dependent on frequency [bearing in mind that the frequency $f = \omega/(2\pi)$] and on the damping ratio q as shown in figure 4.3.

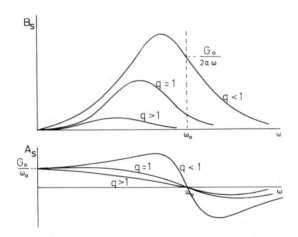

Figure 4.3 The dependences on frequency $f [= \omega/(2\pi)]$ of the harmonic driving force of the amplitudes of the cosine term A_s and of the sine term B_s in the steady-state displacement of the driven mass.

In an underdamped system, $q < 1$, A_s is positive at frequencies less than f_0, the undamped natural frequency of the system, and goes to a peak at $\omega = \sqrt{(\omega_0^2 - 2\alpha\omega_0)}$. At very low frequencies, $A_s = G_0/\omega_0^2$. At frequencies greater than f_0, A_s is negative going to a trough at $\omega = \sqrt{(\omega_0^2 + 2\alpha\omega_0)}$ and tending asymptotically to zero at very high frequencies. On the other hand B_s remains positive at all frequencies tending to zero at high and low frequencies and resonating or going to a peak at some frequency $f < f_0$.

Under the more severe damping conditions of critical damping ($q = 1$) and overdamping ($q > 1$), the magnitude of the A_s term is smaller at most frequencies and it does not go through a peak at a frequency less than f_0. A_s goes to zero at a lower frequency, the greater the damping ratio. At higher damping ratios, the resonance of the other amplitude term, the out-of-phase B_s term, becomes less and less as the damping is increased. Also, the frequency at which the peak occurs becomes lower and lower but always less than f_0 [$= \omega_0/(2\pi)$].

The alternative way of expressing this steady-state response, that of equation 4.8, has parameters C_s and γ_s, thus

$$C_s = \left(\frac{1}{\sqrt{[(\omega_0^2 - \omega^2)^2 + (2\alpha\omega)^2]}}\right) G_0 \tag{4.18}$$

and

$$\gamma_s = \tan^{-1}\left(\frac{2\alpha\omega}{\omega_0^2 - \omega^2}\right). \tag{4.19}$$

This formulation of the steady-state response has the response of the same frequency as the driving force, with an amplitude C_s, and lagging the driving force by a phase angle γ_s. Both of these parameters are dependent on frequency and on damping ratio as shown in figure 4.4.

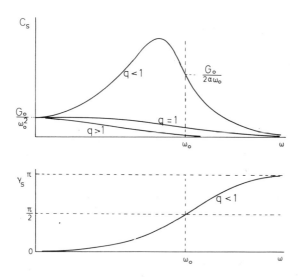

Figure 4.4 The dependences on frequency f [$= \omega/(2\pi)$] of the harmonic driving force, of the amplitude C_s and of the phase angle γ_s of the steady-state displacement of the driven mass.

In an underdamped system for which $q < 1$, the amplitude C_s is always positive, starting at a value G_0/ω_0^2 at low frequency, resonating or going to a maximum at $\omega = \sqrt{(\omega_0^2 - 2\alpha^2)}$ and tending to zero at high frequencies. The maximum value of C_s may be written in a number of ways, using equations 3.37 and 3.54, thus

$$C_s(\max) = \frac{G_0}{2\alpha\sqrt{(\omega_0^2 - \alpha^2)}}$$
$$= \frac{G_0}{2q\omega_0^2\sqrt{(1 - q^2)}}$$
$$= \frac{G_0}{2\alpha^2\sqrt{(4Q^2 - 1)}} \ . \qquad (4.20)$$

Note that as the amplitude undergoes this resonance, the phase angle γ_s rises from zero at low frequencies, through $\pi/2$ at f_0, and approaches π asymptotically at very high frequencies. From equation 4.20 the maximum or resonant displacement amplitude is directly proportional to the amplitude of the driving force F_0 and inversely proportional to the mass M. It also increases as the amount of damping is reduced.

As the damping is increased to critical damping ($q = 1$), the displacement amplitude C_s has no maximum other than its value at low frequencies. No resonance occurs. The phase angle follows the same pattern as for underdamping. In an overdamped system the low frequency amplitude declines to zero at relatively low frequencies and again no resonance occurs.

It should be remembered of course that the total response is the sum of the transient and steady-state responses

$$x = x_t + x_s \qquad (4.21)$$

The transient response is confined in time to soon after start-up while the steady-state response persists for as long as the driving force persists. When this force is stopped, there ensues a switch-off transient similar in type to that at start-up. The initial conditions for this transient would issue from the state of the system at switch-off.

Example 4.1

A machine of mass 100 kg rests on a vertical spring the other end of which stands on the rigid floor. The machine is subjected to an oscillating vertical force of $10 \cos 30t$ N. The damping ratio is 0.1. Before the external force is applied the spring is shortened by 0.01 m. Take the acceleration due to gravity to be 10 m/s^2 and derive the steady-state response of the system.

SINUSOIDAL DRIVING OF THE DAMPED HARMONIC OSCILLATOR

Solution

To find the spring constant, C, note that $Mg = Cx$ where x is the resting deflection 0.01 m. Therefore $C = 10^5$ N/m. So $\omega_0 = \sqrt{(C/M)} = 63.2$ rad/s and $b = 0.1\, b_c = 0.1 \times 2\sqrt{(CM)} = 632$ kg/s. $2\alpha = b/M = 6.32$ 1/s.

Now the complete driving force is $f(t) = -Mg + 10 \cos 30t$ and the response will take the form $x = -0.01 + x_s$, where

$$x_s = C_s \cos(30t - \gamma_s).$$

From equation 4.19

$$\gamma_s = \tan^{-1}(0.06) = 0.12 \text{ rad}.$$

From equation 4.18

$$C_s = 0.1/3100 = 32 \times 10^{-6} \text{ m}.$$

So

$$x = -0.01 + 32 \cos(30t + 0.12) \times 10^{-6} \text{ m}$$

4.3 Mechanical Impedance and Compliance

The mechanical impedance Z_m is defined as the ratio of the driving force to the resulting velocity of the mass. In the situation of figure 4.1, the impedance at the point of application of the force is called the driving point impedance.

$$Z_m = \frac{f(t)}{\dot{x}_s}. \tag{4.22}$$

and has units of kilograms per second (kg/s).

For a thorough analysis of Z_m it is fruitful to consider the complete exponential form of the steady-state displacement, that is, from equation 4.8

$$x_s = C_s e^{j(\omega t - \gamma_s)} \tag{4.23}$$

of which the real part is the solution of equation 4.8. The complex mass velocity then is

$$\dot{x}_s = j\omega C_s e^{j(\omega t - \gamma_s)}. \tag{4.24}$$

The general exponential form for the driving force is

$$f(t) = F_0 \, e^{j\omega t} \tag{4.25}$$

of which the real part is the actual driving force, considered in section 4.2 of this chapter. Now the complex mechanical impedance may be written

$$Z_m = \frac{F_0 \, e^{j\omega t}}{j\omega C_s \, e^{j(\omega t - \gamma_s)}}$$

$$= \left(\frac{F_0}{C_s}\right)\left(\frac{1}{j\omega}\right) e^{j\gamma_s} . \tag{4.26}$$

The mechanical impedance Z_m is therefore a complex number with real and imaginary parts, thus

$$Z_m = \frac{F_0}{\omega C_s} (\sin \gamma_s - j \cos \gamma_s) . \tag{4.27}$$

It can be shown from equations 4.5, 4.18 and 4.19, that

$$\cos \gamma_s = \frac{MC_s}{F_0} (\omega_0^2 - \omega^2) \tag{4.28}$$

and

$$\sin \gamma_s = \frac{MC_s}{F_0} (2\alpha\omega) . \tag{4.29}$$

So

$$Z_m = \frac{M}{\omega} [2\alpha\omega - j(\omega_0^2 - \omega^2)] . \tag{4.30}$$

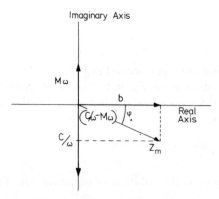

Figure 4.5 The real part b and the imaginary parts $M\omega$ and $-C/\omega$ of the mechanical impedance of the damped harmonic oscillator combine as shown to constitute the complex impedance Z_m.

If the basic system parameters are substituted in equation 4.30, ($\omega_0^2 = C/M$ and $2\alpha = b/M$), then

$$Z_m = b - j\left(\frac{C}{\omega} - M\omega\right)$$

$$= b + j\left(M\omega - \frac{C}{\omega}\right). \qquad (4.31)$$

The components of Z_m may be depicted in the complex plane as shown in figure 4.5. The magnitude of Z_m is clearly given by

$$|Z_m| = \sqrt{\left[b^2 + \left(\frac{C}{\omega} - M\omega\right)^2\right]} \qquad (4.32)$$

and the phase angle of Z_m, φ, is

$$\varphi = \tan^{-1}\left(\frac{C/\omega - M\omega}{b}\right)$$

$$= \frac{\pi}{2} - \gamma_s. \qquad (4.33)$$

The complex mechanical impedance Z_m may therefore be expressed

$$Z_m = |Z_m|\, e^{-j\varphi}. \qquad (4.34)$$

The imaginary part of Z_m is strongly frequency dependent, as illustrated in figure 4.6. The energy storage elements in the system, the mass and the spring, determine this frequency dependence. Note in particular that when

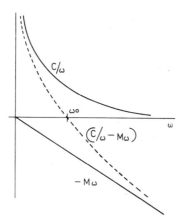

Figure 4.6 The dependences on frequency $f\,[=\omega/(2\pi)]$ of the two imaginary parts $M\omega$ and C/ω of the mechanical impedance. The dependence of their difference $[(C/\omega) - M\omega]$ on frequency is also shown.

$f = f_0$, the undamped natural frequency of the oscillator, the imaginary part of the impedance is zero. Then the impedance is effectively the real part b and the magnitude of Z_m is a minimum. For a fixed magnitude of driving force, the magnitude of the velocity is a maximum at this frequency.

At frequencies less than f_0, the spring part of the imaginary impedance predominates over the mass. At frequencies greater than f_0, the converse holds.

The inverse of the mechanical impedance is called the compliance, Y_m where

$$Y_m = \frac{1}{Z_m}$$

$$= \frac{1}{|Z_m|} e^{j\varphi} \qquad (4.35)$$

Example 4.2

In the oscillator of figure 4.1 the mass is 40 kg, the spring constant is 100 N/m and the damping ratio 0.1. Find the magnitude and phase of the driving point impedance at frequencies 0.1, 1 and 10 times the undamped natural frequency.

Solution

$$\omega_0 = \sqrt{(C/M)} = 1.58 \text{ rad/s} .$$
$$b = 0.1 \, b_c = 0.2 \sqrt{(CM)} = 12.65 \text{ kg/s} .$$

$$Z_m = b + j\left(\frac{C}{\omega} - M\omega\right) .$$

(a) At $\omega = 0.1 \, \omega_0 = 0.16$ rad/s ,

$$Z_m = 12.65 + j(631)$$
$$= 631 \, e^{j1.55} \text{ kg/s} .$$

(b) At $\omega = \omega_0 = 1.58$ rad/s ,

$$Z_m = 12.65 \text{ kg/s}.$$

(c) At $\omega = 10\,\omega_0 = 15.8$ kg/s ,

$$Z_m = 12.65 + j(626)$$
$$= 626\,e^{-j1.55}\text{ kg/s} .$$

4.4 Energy and Power in the Driven Oscillator

In the steady state, the driving force does work on the oscillator. Some of this work is temporarily stored as potential or kinetic energy and the rest is dissipated as heat. The instantaneous rate at which work (Wk) is done on the system is the rate at which energy is given to the system. This is also the instantaneous power transfer to the system $p(t)$

$$p(t) = \dot{W}k$$
$$= f(t)\,\dot{x}_s . \tag{4.36}$$

Therefore, using equations 4.4 and 4.11

$$p(t) = \omega F_0 B_s \cos^2 \omega t - \omega F_0 A_s \cos \omega t \sin \omega t. \tag{4.37}$$

The time-average power P may be calculated by integrating each side of this latter equation over one period T of the input force and then dividing by T, the period. Since

$$\frac{1}{T}\int_t^{t+T} \cos^2 \omega t \, dt = \tfrac{1}{2}$$

(see equation 5.8) and

$$\frac{1}{T}\int_t^{t+T} \cos \omega t \sin \omega t \, dt = 0$$

(see equation 5.4) then

$$\frac{1}{T}\int_t^{t+T} p(t)\,dt = P$$
$$= \tfrac{1}{2}\,\omega F_0 B_s . \tag{4.38}$$

Note that it is the term involving the amplitude B_s in the displacement which is $\pi/2$ out of phase with the driving force which results in power dissipation. B_s is therefore sometimes called the absorptive amplitude of the displacement response. On the other hand, the in-phase part of the displacement x_s, with amplitude A_s, accounts for the temporarily stored energy in the system. A_s is called the elastic or dispersive amplitude of the response. It contributes to the instantaneous power intake of the system but not to the time-averaged power absorption.

It can be shown that P, the time-averaged power absorption given in equation 4.38, is the power dissipated by the damping mechanism. The instantaneous power used to overcome damping is

$$p_d(t) = f_d(t)\dot{x}_s$$
$$= 2\alpha M \dot{x}_s^2 \qquad (4.39)$$

where $f_d(t)$ is the force exerted on the damper $(b\dot{x})$. The time-averaged power to overcome damping P_d is then derived

$$\frac{1}{T}\int_t^{t+T} p_d(t)\,dt = P_d$$

$$= \frac{2\alpha M}{T}\int_t^{t+T} \dot{x}_s^2\,dt$$

$$= \frac{2\alpha\omega^2 M}{T}\int_t^{t+T} (A_s \sin\omega t - B_s \cos\omega t)^2\,dt$$

$$= \frac{2\alpha\omega^2 M}{2}(A_s^2 + B_s^2)\,, \qquad (4.40)$$

since (from equation 5.7)

$$\frac{1}{T}\int_t^{t+T} \sin^2\omega t\,dt = \tfrac{1}{2}\,.$$

Using equations 4.16 and 4.17, which provide expressions for A_s and B_s, equation 4.40 may be clarified thus

$$P_d = \frac{2\alpha\omega^2 M}{2}(A_s^2 + B_s^2)$$
$$= \tfrac{1}{2}\,\omega\,B_s\,F_0$$
$$= P\,. \qquad (4.41)$$

All of the average power input goes to overcome damping.

SINUSOIDAL DRIVING OF THE DAMPED HARMONIC OSCILLATOR

The average power may also be formulated in terms of mechanical impedance quantities. Initially substitute the expression for B_s of equation 4.17 into equation 4.38

$$P = \tfrac{1}{2} \frac{2\alpha\omega^2 F_0^2}{M[(\omega_0^2 - \omega^2)^2 + (2\alpha\omega)^2]} \quad . \tag{4.42}$$

Then substituting the basic system parameters for ω_0^2 and 2α,

$$P = \frac{\tfrac{1}{2} b F_0^2}{b^2 + (C/\omega - M\omega)^2} \tag{4.43}$$

$$= \tfrac{1}{2} \frac{b F_0^2}{|Z_m|^2} \tag{4.44}$$

from equation 4.32. If the concept of the root-mean-square value of the force F is introduced where

$$F = \frac{F_0}{\sqrt{2}}, \tag{4.45}$$

then

$$P = \frac{b F^2}{|Z_m|^2} \quad . \tag{4.46}$$

But $F/|Z_m|$ is the root-mean-square value of the velocity of the mass $\dot{X}$ and so equation 4.46 then becomes

$$P = b\dot{X}^2 \quad . \tag{4.47}$$

Furthermore, from figure 4.5 it can be seen that

$$b = |Z_m| \cos \varphi \tag{4.48}$$

and so one further version of equation 4.38, expressing the time-averaged power dissipation by the driven oscillator, emerges

$$P = F\dot{X} \cos \varphi \quad . \tag{4.49}$$

The factor $\cos \varphi$ is known as the power factor of the system. Note that

$$\cos \varphi = \frac{b}{\sqrt{[b^2 + (C/\omega - M\omega)^2]}} \tag{4.50}$$

and is a strong function of frequency as illustrated in figure 4.7. At low frequencies, $\cos \varphi$ is close to zero as it also is at high frequencies. At $\omega = \omega_0$, $\cos \varphi$ becomes unity.

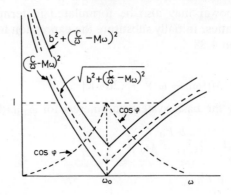

Figure 4.7 Dependence on frequency of the magnitude of the mechanical impedance $|Z_m|$. The variation of the power factor $\cos \varphi$ with frequency is also shown.

Example 4.3

A force of $100 \cos 150\, t$ N horizontally on a mass of 50 kg which in turn is attached to a spring of constant 250 N/m. The damping factor of the oscillator is 0.1. Find the power dissipation in the system and the root-mean-square velocity of the mass.

Solution

$b = 0.1\, b_c = 0.2 \sqrt{(CM)} = 22.4$ kg/s .

Therefore from equation 4.31 the over-all impedance Z_m is

$$Z_m = 22.4 + j(50 \times 150 - 250/150)$$
$$= 22.4 + j\,7500 \text{ kg/s} .$$

Then

$$|Z_m| = 7500 \text{ kg/s} .$$

But the root-mean-square force $F = 100/\sqrt{2} = 71$ N and so the root-mean-square velocity $\dot{X} = 71/7500 = 0.0095$ m/s. The power factor $\cos \varphi = 22.4/7500 = 0.003$ (from equation 4.50). So the power dissipation

$$P = 71 \times 0.0095 \times 0.003 = 2 \text{ mW} .$$

4.5 Power Resonance

It emerges clearly from the foregoing discussion and from figures 4.3, 4.4, 4.6 and 4.7 that the over-all behaviour of the driven harmonic oscillator in the steady state is strongly conditioned by the driving frequency and the degree of damping present. In particular, consider the dependence of the time-averaged power input P on frequency. From equation 4.42

$$P = \frac{F_0^2}{4\alpha M}\left[\frac{(2\alpha\omega)^2}{(\omega_0^2 - \omega^2)^2 + (2\alpha\omega)^2}\right]$$

$$= P_0\left[\frac{(2\alpha\omega)^2}{(\omega_0^2 - \omega^2)^2 + (2\alpha\omega)^2}\right] \qquad (4.51)$$

where P_0 is the value of P at $\omega = \omega_0$. This function P is plotted against ω in figure 4.8. The frequency of undamped free oscillations in the system $f_0 \,[= \omega_0/(2\pi)]$ is also the frequency at which the power intake is a maximum and is therefore also called the power resonant frequency. Power resonance is the phenomenon whereby the power intake is maximised at a particular frequency.

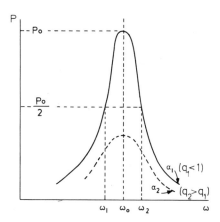

Figure 4.8 Average power dissipated in the damped harmonic oscillator as a function of driving frequency f $[= \omega/(2\pi)]$. This is the power resonance curve of the system. The half-power points are shown. When the damping is increased such that the new decay constant $\alpha_2 > \alpha_1$, the maximum of the curve is reduced and the width between the half-power points increased.

The extent of the resonance or the value of P_0 depends on the square of the amplitude of the driving force and inversely on the damping coefficient

$$P_0 = \frac{F_0^2}{4\alpha M}$$

$$= \frac{F^2}{b} \qquad (4.52)$$

where F is the root-mean-square value of the driving force. For a given driving force amplitude, the greater the damping coefficient present, the lower the resonance peak.

The second feature of significance in the resonance curve of figure 4.8 is the width of the curve. By convention, this is measured as the frequency bandwidth between the two points on the curve at which the power dissipation is one half of the maximum, the so-called half-power points. These two points are often called the 3 dB down points, since on the dB scale, $0.5 = -3$ dB.

The two relevant values of ω are ω_1 and ω_2 and they may be calculated from equation 4.51 as follows:

$$(2\alpha\omega)^2 = 0.5\,[(\omega_0^2 - \omega^2)^2 + (2\alpha\omega)^2] \qquad (4.53)$$
$$(2\alpha\omega)^2 = (\omega_0^2 - \omega^2)^2$$
$$\omega_0^2 - \omega^2 = \pm\, 2\alpha\omega$$

and

$$\omega^2 \pm 2\alpha\omega - \omega_0^2 = 0\,. \qquad (4.54)$$

The solutions to this quadratic equation in ω may be written

$$\omega = -\alpha \pm \sqrt{(\omega_0^2 + \alpha^2)} \qquad (4.55)$$

and

$$\omega = +\alpha \pm \sqrt{(\omega_0^2 + \alpha^2)}\,. \qquad (4.56)$$

Since ω must be positive, the two allowable solutions are

$$\omega_1 = \sqrt{(\omega_0^2 + \alpha^2)} - \alpha \qquad (4.57)$$

and

$$\omega_2 = \sqrt{(\omega_0^2 + \alpha^2)} + \alpha\,. \qquad (4.58)$$

The span from ω_1 to ω_2, $\Delta\omega$, is called the resonance width

$$\Delta\omega = \omega_2 - \omega_1$$
$$= 2\pi\Delta f$$
$$= 2\alpha \ . \tag{4.59}$$

The resonance width is thus directly proportional to the damping coefficient or damping ratio. The resonance frequency divided by the resonance bandwidth Δf is another expression, together with equations 3.53, 3.54, 3.55 and 3.56, for the quality factor of the system

$$\frac{f_0}{\Delta f} = \frac{\omega_0}{\Delta\omega}$$
$$= \frac{\omega_0}{2\alpha} \tag{4.60}$$
$$= Q \ . \tag{4.61}$$

This relationship allows the calculation of the quality value Q from a measured resonance curve such as in figure 4.8. Conversely, knowledge of the value of Q and of the resonance frequency f_0 allows one to predict the shape of the resonance curve, the resonant behaviour of the system. A high value of Q correlates with a peaked resonance curve, a sharp resonance, of a system with little damping.

A further valuable insight into resonance may be gained from a review of the impedance at resonance. From figure 4.6, when

$$\omega = \omega_0$$
$$\frac{C}{\omega_0} - M\omega_0 = 0 \tag{4.62}$$

and

$$Z_m = b \ . \tag{4.63}$$

Note that Z_m and b have the same units, kg/s. The total impedance is accounted for by the damping coefficient b and is therefore real. This does not mean that the mass and the spring do not receive instantaneous power. In fact they do and they may each receive high levels of instantaneous power during each period but at every instant the power being supplied to one of these elements is equal to the power being returned from the other. For example, the instantaneous power to the spring $p_s(t)$ is

$$p_s(t) = Cx_s\dot{x}_s$$
$$= -CC_s^2\omega \cos(\omega t - \gamma_s) \sin(\omega t - \gamma_s) \qquad (4.64)$$

while the instantaneous power to the mass $p_m(t)$ is

$$p_m(t) = M\ddot{x}_s\dot{x}_s$$
$$= MC_s^2\omega^3 \cos(\omega t - \gamma_s) \sin(\omega t - \gamma_s) \ . \qquad (4.65)$$

At $\omega = \omega_0 = \sqrt{(C/M)}$,

$$p_s(t) = -p_m(t) \ . \qquad (4.66)$$

Invoking equation 4.38, one can show that the time-averaged value of each of these powers, P_s and P_m respectively, is zero.

$$P_s = 0 = P_m \ . \qquad (4.67)$$

At resonance, the instantaneous force on the spring is equal to and opposite to the instantaneous force on the mass, thus

$$CC_s \cos(\omega t - \gamma_s) = -M\omega_0^2 C_s \cos(\omega t - \gamma_s) \qquad (4.68)$$

Thus all of the externally applied force is experienced by the viscous damping element

$$f(t) = F_0 \cos \omega_0 t$$
$$= -b\omega_0 C_s \sin(\omega_0 t - \gamma_s) \ . \qquad (4.69)$$

But inspection of equations 4.10, 4.16 and 4.17 shows that at

$$\omega = \omega_0$$

$$\tan \gamma_s = \infty$$

and

$$\gamma_s = \frac{\pi}{2} \ . \qquad (4.70)$$

Therefore, since $\sin(\omega t - \pi/2) = \cos \omega t$,

$$f(t) = -b\omega_0 C_s \cos \omega_0 t \ . \qquad (4.71)$$

SINUSOIDAL DRIVING OF THE DAMPED HARMONIC OSCILLATOR

So the magnitude of C_s is

$$|C_s| = \frac{F_0}{b\omega_0} \tag{4.72}$$

as could also have been derived from equation 4.18.

The ratio of the amplitude of the spring force CC_s to the amplitude of the external force F_0 or $b\omega_0 C_s$ is called the force amplification factor of the oscillator and it may be written down using equation 4.68 and 4.69

$$\frac{CC_s}{b\omega_0 C_s} = \frac{\omega_0}{2\alpha} \tag{4.73}$$

$$= Q \tag{4.74}$$

from equation 3.54. Thus the force magnification factor is identical with the quality factor of the system. In a high Q system the amplitudes of the spring and mass forces are much larger than that of the externally applied force.

Often the equation for the power resonance curve in figure 4.8, equation 4.51, is cited in terms of frequency f and quality factor Q instead of ω and α. Thus, taking $2\alpha = \omega_0/Q$ and $\omega = 2\pi f$, equation 4.51 becomes

$$P = P_0 \left[\frac{(f/f_0 Q)^2}{[1 - (f/f_0)^2]^2 + (f/f_0 Q)^2} \right] \tag{4.75}$$

$$= P_0 \left[\frac{(r/Q)^2}{(1 - r^2)^2 + (r/Q)^2} \right] \tag{4.76}$$

where r is the frequency normalised to the resonant frequency f_0 thus

$$r = \frac{f}{f_0}$$

$$= \frac{\omega}{\omega_0}. \tag{4.77}$$

Example 4.4

A damped mechanical oscillatory system such as that of figure 4.1 is subjected to a force of 50 cos ωt N. When ω = 10 rad/s, the root-mean-square velocity of the mass is a maximum of 0.1 m/s. The mass is 100 kg. Find the Q value of the system and the half-power frequencies.

Solution

$|\dot{X}|$ is maximum at the resonant frequency so $\omega_0 = 10$ rad/s. Then, from equations 4.22 and 4.45

$$|\dot{X}| = 50/\sqrt{(2)}b = 0.1 \text{ and so } b = 354 \text{ kg/s} .$$

Since $\omega_0^2 = C/M = 100 = C/100$, $C = 10^4$ N/m. Now the quality factor $Q = \sqrt{(CM)}/b = 2.8$. Now $f_0 = \omega_0/2\pi = 1.59$ Hz. But $\Delta f = f_0/Q = 0.57$ so

$$f_1 = 1.31 \text{ Hz and } f_2 = 1.88 \text{ Hz} .$$

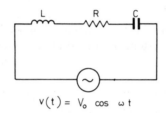

Figure 4.9 The series RLC electrical circuit, supplied by an alternating voltage source, is the direct analogue of the driven damped harmonic oscillator of figure 4.1.

4.6 Electrical Analogue Circuits

The sinusoidally (A.C.) driven RLC circuit of figure 4.9 is the exact analogue of the mechanical system of figure 4.1. The single frequency voltage output of the signal generator

$$v(t) = V_0 \cos \omega t \tag{4.78}$$

is analogous to $f(t)$ in equation 4.4. Otherwise, the analogue is identical with that discussed in section 3.7. In the electrical circuit, the power resonant frquency

$$f_0 = \frac{\omega_0}{2\pi}$$

$$= \frac{1}{2\pi\sqrt{(LC)}} . \tag{4.79}$$

At that frequency, the over-all series electrical impedance

$$Z = R \, \Omega . \tag{4.80}$$

Off resonance, the impedance is

$$Z = R - j\left(\frac{1}{C\omega} - L\omega\right) \tag{4.81}$$

and the power factor $\cos \varphi$ is

$$\cos \varphi = \frac{R}{|Z|} \tag{4.82}$$

The Q value for the circuit may be readily calculated by measuring the voltage magnification factor relationship. By monitoring the ratio of the root-mean-square voltage across the capacitor to the voltage supplied to the circuit as a whole, as a function of frequency, it is found that at $f = f_0$, the ratio is a maximum, Q.

The analogue computer system of figure 3.16 may also be readily used to simulate the driven harmonic oscillator. In this application the alternating voltage $V_0 \cos \omega t$ is first inverted to $-V_0 \cos \omega t$ and then supplied to a potential divider. The output of this potential divider is set at $-(V_0/M)\cos \omega t$ and is fed to the summing input of the final inverter (operational amplifier number 3). Then the feedback line carries the voltage, $-(C/M)q - (b/M)\dot{q} + (V_0/M) \cos \omega t$, which is identical with the analogue of acceleration $\ddot{q}$ from equation 4.2 where the appropriate analogous quantities listed in table 2.1, are used.

4.7 Experiments

(a) Harmonically Driven Damped Oscillator

An electrically driven vibrator, the frequency of which is variable, may be used as the source of the harmonic force to drive a damped mass/spring system such as that in figure 4.1. The amplitude of the motion of the mass, C_s, may be readily measured on the horizontal support surface. To measure the phase angle γ_s an electrical method of simultaneously detecting the instantaneous driving force and the instantaneous mass displacement is needed. A vibrometer mounted on the mass gives a voltage output proportional to the displacement of the mass. A rigid rod, on which strain gauges are mounted along the length, interposed between the vibrator and the mass, produces a voltage proportional to the force. Comparison of these two voltages allows the phase angle to be derived.

If the vibrometer output is differentiated, the result is the velocity of the mass. Knowledge of the instantaneous force and instantaneous velocity allows the mechanical impedance to be completely specified as in equations 4.32 and 4.33.

If the frequency is varied, the amplitude of the velocity is a maximum at power resonant frequency. At this frequency, the net force on the mass (the acceleration of the mass, as measured with an accelerometer multiplied by the mass) is in magnitude equal to the spring force (which may be measured with a force transducer in series with the spring). The ratio of one of these forces at the resonant frequency to the driving force is the Q-value of the system.

(b) Driven *RLC* Electrical Circuit

Experiments similar to the above can be very readily carried out on an electrical *RLC* circuit supplied by an A.C. signal generator. Many functional parameters of the circuit may be derived such as the power resonant frequency when the current is maximum, the Q-value from the ratio of capacitance voltage to input voltage at the resonant frequency, the value of R from the impedance at resonance, the value of L (if C is known) from the value of the resonant frequency, etc.

Problems

4.1 A harmonic oscillator consists of a mass of 4 kg attached to a spring of constant 16 N/m with an associated damping coefficient of 8 kg/s. Motion is horizontal. Find the power resonant frequency, the quality factor and the half-power frequencies.

4.2 Find the magnitude and phase angle of the mechanical impedance of the system of problem 4.1 at a frquency of 0.2 Hz.

4.3 A force of r.m.s. 10 N and of variable frequency is applied to the system of the previous problem. Calculate the r.m.s. velocity of the mass at 0.1, 0.3 and 1 Hz.

4.4 A horizontally moving harmonic oscillator is driven by a sinusoidal force and is observed to have a displacement in the steady state thus, $-0.095 \cos 10t + 0.005 \sin 10t$ m. The mass is 5 kg and the spring constant 45 N/m. Find the decay constant, the damping coefficient and the Q-value for the system. Write down the expression for the driving force.

4.5 Draw an electrical circuit to simulate the set-up of the previous problem and calculate the average power dissipation in the system. What would the maximum power be at resonance?

SINUSOIDAL DRIVING OF THE DAMPED HARMONIC OSCILLATOR

4.6 Show that the velocity of the mass in the driven damped harmonic system is maximum at the power resonance frequency.

4.7 Calculate the frequency at which the steady-state acceleration amplitude of the mass is maximum and find the value of that maximum acceleration.

4.8 A horizontally moving harmonic oscillator consists of a mass of 1 kg, a spring of constant 15 N/m and a damper of coefficient 5 kg/s. Find the Q-value of the system. It is acted upon by a horizontal force $\cos t$ N. Find the magnitude and phase angle of the impedance at the frequency of the driving force. Calculate the amplitude of the displacement, that of the velocity and the average power dissipated.

4.9 If the driving force in the system of the previous problem has the same amplitude and the frequency at which the system resonates, then find the average power, the velocity amplitude and displacement amplitude.

4.10 An electrical circuit like that of figure 4.9 has a capacitor of 10^{-8} F capacitance and an inductor coil of 10^{-6} H inductance. The Q-value is 50. The signal generator has an output of amplitude 10 V and variable frequency. Find the resistance in the circuit, the power resonance frequency and the power (average) and current amplitude at that frequency.

4.11 Draw a graph of the magnitude and phase angle of the impedance of the circuit in the previous problem as functions of frequency over the two decades above and below the resonant frequency.

4.12 A driving force at 1 Hz acts on an overdamped system of decay constant 0.5 1/s and produces a steady-state displacement amplitude of 5 cm. If the frequency is changed to 10 Hz while the force amplitude is kept constant, write down the complete expression for the new displacement.

4.13 A moving-coil loudspeaker may be viewed as a damped harmonic oscillator which is driven by the current in the coil. The following data are given about such a loudspeaker: mass of coil and diaphragm, 0.02 kg; spring constant of coil/diaphragm mounting, 2.5 kN/m; damping coefficient, 5 kg/s; amplitude of force, 5 N. Find the amplitude of the displacement of the coil at 1, 100 and 10000 Hz. What is the mechanical Q-value, the resonant frequency and the half-power frequencies for the loudspeaker. The effects of the air loading on the diaphragm may be neglected.

Further Reading

Anderson, R.A., *Fundamentals of Vibrations* (Macmillan, New York, 1967)
Bishop, R.E.D., *Vibration* (Cambridge University Press, Cambridge, 1965)
Braddick, H.J.H., *Vibrations, Waves and Diffraction* (McGraw-Hill, London, 1965)
Church, A.H., *Mechanical Vibrations*, 2nd edition (Wiley, New York, 1963)
Clayton, G.B., *Operational Amplifiers* (Butterworths, London, 1974)
Daish, C.B. and Fender, D.H., *Experimental Physics*, 2nd edition (Hodder & Stoughton, London, 1976)
Dimarogonas, A.D., *Vibration Engineering* (West, St. Paul, 1976)
French, A.P., *Vibrations and Waves* (Nelson, London, 1971)
Graeme, J.G., *Applications of Operational Amplifiers* (McGraw-Hill, New York, 1974)
Haberman, C.M., *Vibrational Analysis* (Merrill, Columbus, 1968)
Kinsler, L.W. and Frey, A.R., *Fundamentals of Acoustics*, 2nd edition (Wiley, New York, 1962)
Main, I.G., *Vibrations and Waves* (Cambridge University Press, Cambridge, 1978)
Morse, P.M., *Vibration and Sound*, 2nd edition (McGraw-Hill, New York, 1948)
Seto, W.W., *Theory and Problems of Mechanical Vibrations* (Schaum, New York, 1964)
Steidel, R.F. Jr., *An Introduction to Mechanical Vibrations*, 2nd edition (Wiley, New York, 1979)
Timoshenko, S., Young, D.H. and Weaver, W. Jr., *Vibration Problems in Engineering*, 4th edition (Wiley, New York, 1974)
Tyler, F., *A Laboratory Manual of Physics*, 5th edition, SI version (Arnold, London, 1977)
Vernon, J.B., *Linear Vibration Theory* (Wiley, New York, 1967)

5 Harmonic Analysis and Transform Techniques in Damped Harmonic Oscillator Systems

5.1 Introduction and Objectives

In this chapter the aims are to discuss the utility and applications of harmonic analysis and transform techniques in the analysis of damped harmonic oscillators.

After reading this chapter the student should be able to

(a) define the Fourier analysis of a periodic function;
(b) apply the Fourier analysis to a range of tractable periodic functions;
(c) prove that a truncated Fourier series is the best-fit series of harmonics, to a given periodic function, on a least square error basis;
(d) define the Fourier transform and show that an isolated pulse or shock function (non-periodic) can be represented by a continuous spectrum of harmonic components;
(e) define the Laplace transform and derive the basic theorems relating to this transform;
(f) define linearity in relation to driven mechanical oscillatory systems;
(g) outline the application of Laplace transform methods to mechanical oscillatory systems;
(h) define the impulse response and the system transfer function;
(i) use the convolution theorem to calculate the response of the mechanical oscillatory system to an arbitrary driving force.

5.2 Fourier Theory of Harmonic Analysis

The discussion in the previous chapter deals with a driving force which is a purely sinusoidal function. It is rare to have such a pure driving function. Much more common is a non-sinusoidal but repetitive or periodic force function such as that shown in figure 5.1. The period of the function is T s so that

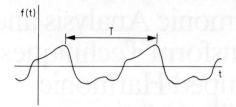

Figure 5.1 A general periodic function of period T.

$$f(t) = f(t - T) \quad (5.1)$$

The Fourier theory states that such a periodic function may be represented by a series expansion of sinusoidal terms consisting of a fundamental frequency and its higher harmonics thus

$$f(t) = a_0 + \sum_{k=1}^{\infty} a_k \cos k\omega_f t + \sum_{k=1}^{\infty} b_k \sin k\omega_f t \quad (5.2)$$

where

$$\omega_f = \frac{2\pi}{T} \quad (5.3)$$

so that the fundamental frequency $f_f = 1/T = \omega_f/2\pi$.

Before calculating the values of the amplitude terms a_0, a_k and b_k, (where $k = 1 \to \infty$) recall the orthogonality of the sine and cosine functions over a period. Thus

$$\int_0^T \sin n\omega_f t \sin m\omega_f t \, dt = 0 \quad n \neq m \quad (5.4)$$

$$\int_0^T \cos n\omega_f t \cos m\omega_f t \, dt = 0 \quad n \neq m \quad (5.5)$$

$$\int_0^T \cos n\omega_f t \sin m\omega_f t \, dt = 0 \quad n \neq m$$

$$= 0 \quad n = m \quad (5.6)$$

$$\int_0^T \sin^2 n\omega_f t \, dt = \frac{T}{2} \quad (5.7)$$

and
$$\int_0^T \cos^2 n\omega_f t \, dt = \frac{T}{2} . \tag{5.8}$$

Using equation 5.2 it can be shown that

$$\int_0^T f(t) \cos k\omega_f t \, dt = \int_0^T a_k \cos^2 k\omega_f t \, dt = a_k \frac{T}{2} \tag{5.9}$$

since, by the orthogonality relations, equations 5.4 to 5.8, all the other terms in the former integral are zero. Hence the amplitude term a_k is given by

$$a_k = \frac{2}{T} \int_0^T f(t) \cos k\omega_f t \, dt . \tag{5.10}$$

Similarly it can be shown that

$$b_k = \frac{2}{T} \int_0^T f(t) \sin k\omega_f t \, dt \tag{5.11}$$

and

$$a_0 = \frac{1}{T} \int_0^T f(t) \, dt . \tag{5.12}$$

Thus this latter term a_0 is the D.C. level of $f(t)$ while a_k and b_k are the amplitudes of the relevant (k-th harmonic) cosine and sine terms in the Fourier series. Equation 5.2 may be reformulated thus

$$f(t) = a_0 + \sum_{k=1}^{\infty} c_k \cos (k\omega_f t - \varphi_k) \tag{5.13}$$

where

$$c_k = \sqrt{(a_k^2 + b_k^2)} \tag{5.14}$$

and

$$\tan \varphi_k = \frac{b_k}{a_k} . \tag{5.15}$$

By extending equation 5.13 the periodic function $f(t)$ may also be represented by an infinite exponential expansion thus

$$f(t) = a_0 + \sum_{k=1}^{\infty} \left(\frac{c_k}{2}\right) [e^{j(k\omega_f t - \varphi_k)} + e^{-j(k\omega_f t - \varphi_k)}]$$

$$= a_0 + \sum_{k=1}^{\infty} \left[\left(\frac{c_k e^{-j\varphi_k}}{2}\right) e^{jk\omega_f t} + \left(\frac{c_k e^{j\varphi_k}}{2}\right) e^{-jk\omega_f t}\right]$$

$$= d_0 + \sum_{k=1}^{\infty} (d_k e^{jk\omega_f t} + d_k^* e^{-jk\omega_f t}) \qquad (5.16)$$

where

$$d_0 = a_0 \qquad (5.17)$$

$$d_k = \frac{c_k e^{-j\varphi_k}}{2} \qquad (5.18)$$

and d_k^* is the complex conjugate of d_k. But also $d_k^* = d_{-k}$ and so equation 5.16 may be compressed into

$$f(t) = \sum_{k=-\infty}^{\infty} d_k e^{jk\omega_f t} \qquad (5.19)$$

where in fact

$$d_k = \frac{1}{T} \int_0^T f(t) e^{-jk\omega_f t} dt . \qquad (5.20)$$

The physically realisable $f(t)$ is the real part of the expansion of equation 5.19. Thus in practice

$$f(t) = \text{Re} \sum_{k=-\infty}^{\infty} d_k e^{jk\omega_f t} . \qquad (5.21)$$

Thus the real function $f(t)$ may be represented along the frequency axis, that is, in the frequency domain, as shown in figure 5.2, as a series of appropriate amplitudes at the sharply defined harmonic frequencies. Such a representation is called a line spectrum, a series of discrete lines in the frequency domain.

Occasionally, the form of $f(t)$ allows simplification in the derivation of the amplitudes or coefficients in the Fourier series. For instance, if $f(t)$ is an even function of t such that

$$f(t) = f(-t) \qquad (5.22)$$

and if equation 5.11 is written thus

$$b_k = \frac{2}{T} \int_{-T/2}^{+T/2} f(t) \sin k\omega_f t \, dt \qquad (5.23)$$

HARMONIC ANALYSIS IN DAMPED HARMONIC OSCILLATOR SYSTEMS

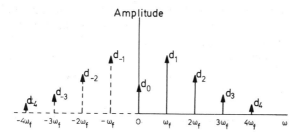

Figure 5.2 The Fourier line spectrum of a periodic real function $f(t)$ showing the amplitudes d_k of each discrete kth harmonic.

by shifting the limits of integration backwards by $T/2$, this integration can be broken up into two segments thus

$$b_k = \frac{2}{T} \left[\int_{-T/2}^{0} f(-t) \sin(-k\omega_f t) \, dt + \int_{0}^{T/2} f(t) \sin(k\omega_f t) \, dt \right].$$

But the product of an even function $f(t)$ and an odd function $\sin(k\omega_f t)$ yields an odd function. Therefore the two integrations in the above equation will yield the same magnitude but will have opposite polarities.

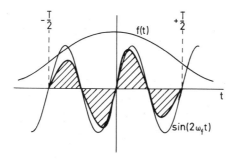

Figure 5.3 The product of an even function $f(t)$ and an odd function $\sin(2\omega_f t)$ produces an odd function which is cross-hatched.

Figure 5.3 illustrates such a process and clearly the cross-hatched area under the product odd function amounts to zero over a period T. Thus in this case

$$b_k = 0 . \tag{5.24}$$

For an even function $f(t)$ there are only cosine terms in the Fourier series expansion together with the D.C. level a_0. All of the phase angles in the expansion of equation 5.13, φ_k, are zero.

If $f(t)$ is odd such that

$$f(t) = -f(-t) \tag{5.25}$$

the amplitudes of the cosine terms a_k may be calculated from equation 5.10

$$a_k = \frac{2}{T} \int_{-T/2}^{+T/2} f(t) \cos(k\omega_f t)\, dt$$

$$= \frac{2}{T} \left[\int_{-T/2}^{0} f(-t) \cos(-k\omega_f t)\, dt + \int_{0}^{T/2} f(t) \cos(k\omega_f t)\, dt \right].$$

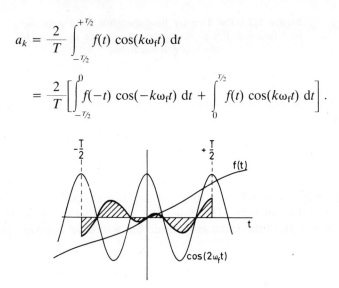

Figure 5.4 The product of an odd function $f(t)$ and a second odd function $\sin(2\omega_f t)$ produces an odd function indicated by the cross-hatching.

As shown in figure 5.4, the product of an odd function $f(t)$ and an even function $\cos(k\omega_f t)$ yields an odd function. The areas under the negative and positive half-periods, shown cross-hatched in the diagram, are equal and of opposite sign. Therefore

$$a_k = 0 \tag{5.26}$$

in the Fourier series expansion for an odd function $f(t)$ and there are only sine terms in the expansion. There is no D.C. level since $a_0 = 0$.

If $f(t)$ is symmetrical about $f(t) = 0$ then there is no D.C. term in the Fourier series expansion or $a_0 = 0$.

The Fourier theory is normally expressed, as in this discussion, as the decomposition of the periodic function $f(t)$ but it should be borne in mind that the reverse process is also valid. If a set of functions of related harmonic frequencies are added together, the aggregate constitutes a periodic function, of period T, equal to the period of the lowest or fundamental harmonic in the constituent functions.

Example 5.1

A periodic triangular wave described over the period from $t = 0$ to $t = T$ by

$$f(t) = -1 + \frac{4}{T} \qquad 0 \leq t \leq \frac{T}{2}$$

$$= 1 - \frac{4}{T}\left(t - \frac{T}{2}\right), \quad \frac{T}{2} \leq t \leq T$$

is to be Fourier analysed. Find the Fourier series coefficients.

Solution

$f(t)$ is even and so $b_k = 0$. Also $f(t)$ is symmetrical about $f(t) = 0$, so $a_0 = 0$. So from equation 5.10.

$$a_k = \frac{2}{T}\int_0^T f(t) \cos k\omega_f t \, dt, \text{ where } \omega_f = \frac{2\pi}{T}.$$

Since both $f(t)$ and $\cos k\omega_f t$ are even, so also is their product. Therefore the integral is the same in each half of the period T. Then

$$a_k = \frac{1}{T}\int_0^{T/2} f(t) \cos\left(\frac{2\pi k t}{T}\right) dt.$$

Changing the variable of integration from t to x $(= 2\pi kt/T)$

$$a_k = \frac{1}{2\pi k}\int_0^{k\pi}\left(\frac{1}{2\pi k} x - 1\right) \cos x \, dx$$

$$= \frac{1}{2\pi k}\left(\frac{1}{2\pi k}\left|\cos x + x \sin x\right|_0^{k\pi} - \left|\sin x\right|_0^{k\pi}\right)$$

$$= -\frac{1}{2\pi^2 k^2} \qquad \text{for } k \text{ odd}$$
$$= 0 \qquad\qquad \text{for } k \text{ even}.$$

There are odd cosine harmonics only and they decline inversely with k^2.

5.3 Truncated Fourier Series

The Fourier series expansions of equations 5.2, 5.13 and 5.19 are infinite. Frequently it is not feasible to consider the infinite series but rather it is sufficient to work with the series truncated at some limited number, K, of harmonics. The truncated series would then be only an approximation to $f(t), f_A(t)$ thus

$$f_A(t) = \sum_{k=0}^{K} \left[a_k \cos(k\omega_f t) + b_k \sin(k\omega_f t) \right] . \tag{5.27}$$

A different truncated series of K harmonics may be postulated to approximately represent $f(t)$, $f_B(t)$ thus

$$f_B(t) = \sum_{k=0}^{K} \left[p_k \cos(k\omega_f t) + q_k \sin(k\omega_f t) \right] \tag{5.28}$$

where p_k and q_k are arbitrary coefficients.

It can be shown that the truncated Fourier series of equation 5.27 provides a better fit to $f(t)$ than any possible version of equation 5.28 on the basis of the least squares criterion. The average square of the error in approximation B is given

$$S = \frac{1}{T} \int_0^T [f(t) - f_B(t)]^2 \, dt \tag{5.29}$$

For S to be a minimum $\partial S/\partial p_k$ must be zero. Thus

$$\frac{\partial S}{\partial p_k} = 0$$

$$= \frac{1}{T} \int_0^T \frac{\partial}{\partial p_k} [f(t) - f_B(t)]^2 \, dt$$

$$= \frac{1}{T} \int_0^T 2[f(t) - f_B(t)] [-\cos(k\omega_f t)] \, dt . \tag{5.30}$$

If in this latter integral the terms are separated one finds

$$\int_0^T f(t) \cos(k\omega_f t) \, dt = \int_0^T f_B(t) \cos(k\omega_f t) \, dt$$

$$= \int_0^T (\Sigma) \cos(k\omega_f t) \, dt \tag{5.31}$$

where Σ denotes the sum of the truncated series of equation 5.28. Using the relevant orthogonality relationships (equations 5.5, 5.6 and 5.8) the right-hand side of equation 5.31 may be simplified to

$$\int_0^T p_k \cos^2(k\omega_f t) \, dt = \frac{p_k T}{2} . \tag{5.32}$$

Therefore, the value of p_k for the least squared error in the approximation

$$p_k = \frac{2}{T} \int_0^T f(t) \cos(k\omega_f t) \, dt$$

$$= a_k \tag{5.33}$$

from equation 5.10.

Likewise it can be shown that, by the same least squared error criterion, the best value of q_k is

$$q_k = b_k \tag{5.34}$$

of equation 5.11.

Therefore the truncated Fourier series is the best fit to $f(t)$ according to the least squared error criterion for a truncated series of the same set of harmonics.

The more terms used in the truncated Fourier series $f_A(t)$, the higher the value of K, the closer the approximation to $f(t)$, provided this function is a smooth, continuous one. This approximation consists of $f_A(t)$ constantly overshooting and undershooting $f(t)$. If $f(t)$ has discontinuities, abrupt changes in value or slope, the Fourier series provides poor approximation, excessive overshooting and undershooting in the vicinity of each discontinuity.

As an example of this latter phenomenon, take the case of a square wave shown in figure 5.5. Over the period T this function may be written

$$f(t) = U(t) \qquad\qquad 0 \leq t \leq \frac{T}{2}$$

$$= U(t) - 2 U(t - \frac{T}{2}) . \qquad \frac{T}{2} \leq t \leq T \tag{5.35}$$

Here $U(t)$ is a unit step at $t = 0$. Alternatively one may write

$$f(t) = 1 \qquad\qquad 0 \leq t \leq \frac{T}{2}$$

$$= -1 . \qquad \frac{T}{2} \leq t \leq T \tag{5.36}$$

This is an odd function of t and it is symmetrical about $f(t) = 0$ so $a_k = 0$ for all values of k. From equation 5.11, b_k may be found thus

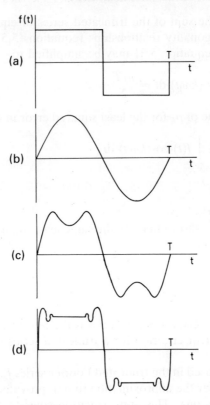

Figure 5.5 The square wave in (a) may be approximated by (b) the first harmonic of the Fourier series expansion, (c) the first and third harmonics and (d) a large number of harmonics in the truncated expansion. On the least squared error basis the approximation improves as more harmonics are added but, regardless of the number of harmonics, there is an overshoot at every discontinuity, the so-called Gibbs' phenomenon.

$$b_k = \frac{2}{T} \int_0^T f(t) \sin(k\omega_f t)\, dt$$

$$= \frac{2}{T} \left[\int_0^{T/2} \sin(k\omega_f t)\, dt - \int_{T/2}^T \sin(k\omega_f t)\, dt \right]$$

$$= \frac{4}{T} \int_0^{T/2} \sin(k\omega_f t)\, dt$$

$$= -\frac{4}{T}\left(\frac{1}{k\omega_f}\right)\cos(k\omega_f t)\Big|_0^{T/2}$$

$$= \frac{2}{k\pi} = \frac{0.64}{k} \quad \text{for } k \text{ odd}$$

$$= 0 \quad \text{for } k \text{ even.} \tag{5.37}$$

There are odd sine harmonics only in the Fourier series expansion to represent the square wave. If the series be truncated at the first harmonic, the approximation shown in figure 5.5(b) is very rough. The addition of the third harmonic improves matters as figure 5.5(c) shows. The more harmonics used, the better the over-all approximation. Even with a large number of harmonics, however, (see figure 5.5(d)) the approximation at each discontinuity has an 18 per cent overshoot. The more harmonics, the more rapid the settling down in the vicinity of the discontinuity but the overshoot is never eliminated. This is Gibbs' phenomenon and it arises from the inability of the Fourier series to accurately represent discontinuities in the periodic function. It is not a truncation error but is inherent in the Fourier series representation.

5.4 Pulse Trains, Isolated Pulses and Shocks

Pulsed forces constitute a very important class of driving functions in mechanical vibratory systems. Consider a train of rectangular pulses, each of amplitude F_0, duration τ and period between pulses T, as shown in

Figure 5.6 A train of rectangular pulses, each of amplitude F_0 and duration τ. The period between the pulses is T.

figure 5.6. For computational convenience take the origin of the t-axis at the mid-point of one pulse. This pulse may be represented thus

$$f(t) = F_0 \left[U\left(t + \frac{\tau}{2}\right) - U\left(t - \frac{\tau}{2}\right) \right] \tag{5.38}$$

and it is repeated every period T. In analysing this function into a Fourier series, it may firstly be noted that it is an even function and therefore all the b_k coefficients are zero. Equation 5.10 may be used to calculate the values of a_k thus

$$a_k = \frac{2}{T} \int_{-T/2}^{+T/2} f(t) \cos(k\omega_f t) \, dt$$

where the integration is taken over the period from $-T/2$ to $+T/2$. Substituting for $f(t)$ from equation 5.38 one may simplify equation 5.38

$$a_k = \frac{2F_0}{T} \int_{-\tau/2}^{+\tau/2} \cos(k\omega_f t) \, dt$$

$$= \frac{2F_0}{T} \left(\frac{1}{k\omega_f}\right) \left| \sin(k\omega_f t) \right|_{-\tau/2}^{\tau/2}$$

$$= \left(\frac{2F_0 \tau}{T}\right) \frac{\sin(k\pi\tau/T)}{(k\pi\tau/T)} \, . \tag{5.39}$$

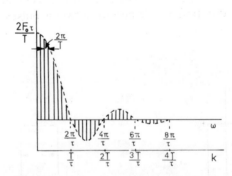

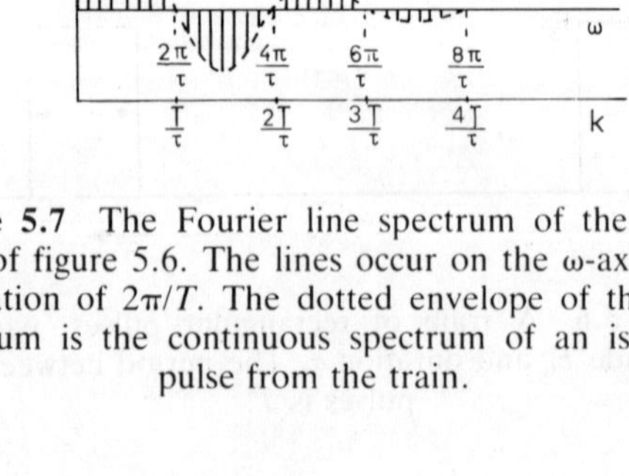

Figure 5.7 The Fourier line spectrum of the pulse train of figure 5.6. The lines occur on the ω-axis at a separation of $2\pi/T$. The dotted envelope of the line spectrum is the continuous spectrum of an isolated pulse from the train.

Figure 5.7 illustrates how this a_k varies with k, the harmonic number, for a given τ/T ratio. Each k corresponds to a Fourier spectral line, to a frequency present in the Fourier series expansion. The separation between the spectral lines on the ω-axis is $2\pi/T$. If the period T is short there are few spectral lines. Conversely, as the period T becomes longer, more and more spectral lines crowd into the ω axis. In the limit of $T \to \infty$, the case of an isolated pulse, the spectral lines in figure 5.7 become so concentrated that the spectrum becomes continuous. All frequencies are then present, to one extent or another, as indicated by the envelope of figure 5.7. This is the continuous Fourier spectrum of the isolated rectangular pulse.

For a given fixed value of the period, T of the pulse train, increasing the pulse duration τ, increases the amplitude of the envelope of figure 5.7 at every frequency in direct proportion. In this case also there are fewer harmonics within each lobe of the envelope. The two converse changes occur when the pulse duration τ is reduced. This is the phenomenon of reciprocal spreading and may be summarised by stating that increasing the pulse duration produces fewer harmonics but at higher amplitudes within the main lobe and vice versa when the pulse duration is decreased.

The isolated rectangular pulse is an example of a shock type of driving force often found to act on mechanical oscillatory systems in real life. If such a shock acts on a mechanical system such as that of figure 4.1 it is very likely that the oscillator will be driven into resonance by the components in the continuous spectrum close to the system resonant frequency. Different types of shock — triangular, pyramidal, semi-circular, etc., in the time domain — have spectra differing in details from that of the rectangular shock but all have continuous Fourier spectra.

The idea of a continuous spectrum for these non-periodic functions leads to the Fourier transform. To develop this transform, recall that the periodic $f(t)$ may be represented

$$f(t) = \sum_{k=-\infty}^{\infty} d_k \, e^{jk\omega_f t} . \tag{5.19}$$

Taking the period from $-\dfrac{T}{2}$ to $+\dfrac{T}{2}$, d_k is given by

$$d_k = \frac{1}{T} \int_{-T/2}^{T/2} f(t) \, e^{-jk\omega_f t} dt . \tag{5.20}$$

If now $T \to \infty$, the frequencies of the spectral components are no longer discrete variables $k\omega_f$ but a continuous variable ω. Then

$$T d_k \xrightarrow[T \to \infty]{} \int_{-\infty}^{\infty} f(t) \, e^{-j\omega t} dt$$

$$= F(\omega) \tag{5.40}$$

where $F(\omega)$ is the Fourier transform or Fourier integral of $f(t)$. Furthermore, the inversion of the Fourier integral may be found from equation 5.19 thus

$$f(t) = \frac{1}{2\pi} \sum_{k=-\infty}^{\infty} T d_k \, e^{jk\omega_f t} \, \omega_f$$

$$\xrightarrow[T\to\infty]{} \frac{1}{2\pi} \int_{-\infty}^{\infty} F(\omega) e^{j\omega t} \, d\omega \qquad (5.41)$$

if the sum is taken in the limit to be an integral and, as $T\to\infty$, ω_f becomes infinitesimal, $d\omega$. Equation 5.41 is therefore the inverse Fourier transform of $F(\omega)$.

A related transform which has come to be more widely used than the Fourier transform in the analysis of linear mechanical and electrical systems, is the Laplace transform.

Example 5.2

Find the Fourier transform of an isolated triangular pulse function of unit area, thus

$$f(t) = \frac{1}{\tau}\left(1 - \frac{|t|}{\tau}\right) \qquad |t| \leq \tau$$

$$= 0 \, . \qquad |t| > \tau$$

Solution

(a) Fourier Series Approach
Assume initially the pulse repeated with period T; the function is even and so $b_k = 0$; also $\omega_f = 2\pi/T$.

$$a_k = \frac{2}{T\tau} \int_0^\tau \left(1 - \frac{t}{\tau}\right) \cos(k\omega_f t) \, dt \, ,$$

and changing the variable to

$$x = k\omega_f t$$

$$a_k = \frac{2}{T\tau k\omega_f} \int_0^{k\omega_f \tau} \left(1 - \frac{x}{k\omega_f \tau}\right) \cos x \, dx$$

$$= \frac{2}{T\tau k\omega_f} \left(\left| \sin x \right|_0^{k\omega_f \tau} - \frac{1}{k\omega_f \tau} \left| \cos x + x \sin x \right|_0^{k\omega_f \tau} \right)$$

HARMONIC ANALYSIS IN DAMPED HARMONIC OSCILLATOR SYSTEMS

$$= \frac{2}{T(k\omega_f\tau)^2} [1 - \cos(k\omega_f\tau)]$$

$$= \frac{1}{T} \left[\frac{\sin(k\omega_f\tau/2)}{(k\omega_f\tau/2)} \right]^2 .$$

And $F(\omega) = Ta_k$ as $T \to \infty$ and $k\omega_f \to \omega$

$$= \left[\frac{\sin(\omega\tau/2)}{(\omega\tau/2)} \right]^2 .$$

(b) From equation 5.40

$$F(\omega) = \int_{-\infty}^{\infty} f(t) \, e^{-j\omega t} dt$$

$$= \frac{1}{\tau} \int_{-\tau}^{0} \left(1 + \frac{t}{\tau}\right) e^{-j\omega t} \, dt + \frac{1}{\tau} \int_{0}^{\tau} \left(1 - \frac{t}{\tau}\right) e^{-j\omega t} \, dt$$

$$= \frac{1}{\tau} \left| -\frac{1}{j\omega} e^{-j\omega t} - \frac{1}{\tau\omega^2} e^{-j\omega t} (-j\omega t - 1) \right|_{-\tau}^{0}$$

$$+ \frac{1}{\tau} \left| -\frac{1}{j\omega} e^{-j\omega t} + \frac{1}{\tau\omega^2} e^{-j\omega t} (-j\omega t - 1) \right|_{0}^{\tau}$$

$$= \frac{1}{\tau} \left[\frac{2}{\tau\omega^2} - \frac{1}{\tau^2\omega^2} (e^{j\omega\tau} + e^{-j\omega\tau}) \right]$$

$$= \frac{2}{\omega^2\tau^2} (1 - \cos\omega\tau)$$

$$= \left[\frac{\sin(\omega\tau/2)}{(\omega\tau/2)} \right]^2 .$$

5.5 Laplace Transforms

In calculating the response of the damped oscillatory system to an arbitrary driving force, $f(t)$, Laplace transforms are very useful. Laplace transforms may in fact be used to solve many simple linear differential equations and because these transforms are tabulated and published, the solutions may be quickly and readily derived. The definition of the Laplace transform $L(s)$ of $f(t)$ is, for an $f(t)$ which is causal, that is, $f(t) = 0$ for $t < 0$

$$L(s) = \int_0^\infty f(t) \, e^{-st} \, dt = \mathscr{L}[f(t)] \tag{5.42}$$

where s is called the transform parameter and $\mathscr{L}$ indicates the Laplace transform operation. In general the inversion process, the derivation of $f(t)$ from $L(s)$, is difficult. In practice however, either experience or a perusal of the published tables of Laplace transforms provides a ready solution. Symbolically the inversion process can be written

$$\mathscr{L}^{-1}[L(s)] = f(t) \ . \tag{5.43}$$

Table 5.1 Laplace Transforms of some Common Functions

$\delta(t)$	1
$U(t)$	$\dfrac{1}{s}$
t^n (for $n = 1, 2, 3\ldots$)	$\dfrac{n!}{s^{n+1}}$
e^{-at}	$\dfrac{1}{s+a}$
$\cos at$	$\dfrac{s}{s^2 + a^2}$
$\sin at$	$\dfrac{a}{s^2 + a^2}$
$t e^{-at}$	$\dfrac{2}{(s+a)^2}$
$t \cos at$	$\dfrac{s^2 - a^2}{(s^2 + a^2)^2}$
$t \sin at$	$\dfrac{2as}{(s^2 + a^2)^2}$
$e^{-bt} \cos at$	$\dfrac{s+b}{(s+b)^2 + a^2}$
$e^{-bt} \sin at$	$\dfrac{a}{(s+b)^2 + a^2}$
$\dot{f}(t)$	$sL(s) - f(0)$
$\ddot{f}(t)$	$s^2 L(s) - sf(0) - \dot{f}(0)$

A number of common functions and their Laplace transforms are listed in table 5.1. More comprehensive tabulations may be found in the references at the end of this chapter. The derivation of each of these transforms follows from equation 5.42. For example the transform of $f(t) = \cos at$, for $t > 0$, may be found thus

$$\mathscr{L}(\cos at) = \int_0^\infty \cos at\, e^{-st}\, dt$$

$$= \frac{a}{s^2 + a^2}\left|-s\cos at + a\sin at\right|_0^\infty$$

$$= \frac{s}{s^2 + a^2}. \tag{5.44}$$

The transform of the step function $f(t) = U(t)$ is found

$$\mathscr{L}[U(t)] = \int_0^\infty U(t)\, e^{-st} dt$$

$$= \int_0^\infty e^{-st} dt$$

$$= -\frac{e^{-st}}{s}\bigg|_0^\infty$$

$$= \frac{1}{s}. \tag{5.45}$$

A number of basic theorems relating to the Laplace transform shed light on its utility in the solution of linear differential equations. They also aid in the inversion process.

(a) Linearity

If $L_1(s)$ and $L_2(s)$ are the Laplace transforms of $f_1(t)$ and $f_2(t)$ respectively and if k_1 and k_2 are arbitrary constants then

$$k_1 L_1(s) + k_2 L_2(s) = \mathscr{L}[k_1 f_1(t) + k_2 f_2(t)]. \tag{5.46}$$

(b) Time Scaling

If k is an arbitrary constant then

$$L\left(\frac{s}{k}\right) = \mathscr{L}[|k|\, f(kt)]. \tag{5.47}$$

(c) Time Shifting
Where t_0 is a constant time shift

$$e^{-st_0}L(s) = \mathscr{L}[f(t - t_0)] \ . \tag{5.48}$$

(d) s-axis Shifting
Where s_0 is a real constant

$$L(s - s_0) = \mathscr{L}[e^{-s_0 t}f(t)] \ . \tag{5.49}$$

(e) Time Differentiation

$$sL(s) - f(0) = \mathscr{L}[f'(t)] \ . \tag{5.50}$$

and

$$s^2 L(s) - sf(0) - f'(0) = \mathscr{L}[f''(t)] \tag{5.51}$$

where $f(0)$ is the initial value of $f(t)$ and $f'(0)$ is the initial value of $f'(t)$.

(f) s Differentiation

$$\frac{dL(s)}{ds} = \mathscr{L}[-tf(t)] \ , \tag{5.52}$$

and

$$\frac{d^2 L(s)}{ds^2} = \mathscr{L}[t^2 f(t)] \ . \tag{5.53}$$

(g) Convolution
If two functions $f_1(t)$ and $f_2(t)$ are given, the convolution of $f_1(t)$ and $f_2(t)$ is defined thus

$$f(t) = \int_{-\infty}^{\infty} f_1(\tau) f_2(t - \tau) \ d\tau \ . \tag{5.54}$$

The convolution procedure is abbreviated symbolically thus

$$f(t) = f_1(t) * f_2(t) \ . \tag{5.55}$$

HARMONIC ANALYSIS IN DAMPED HARMONIC OSCILLATOR SYSTEMS

It can be readily seen that convolution is commutative

$$f_1(t) * f_2(t) = f_2(t) * f_1(t) \ . \tag{5.56}$$

If $L(s)$, $L_1(s)$ and $L_2(s)$ are the Laplace transforms of $f(t)$, $f_1(t)$ and $f_2(t)$ respectively, it can be shown that

$$L(s) = L_1(s) L_2(s) \ . \tag{5.57}$$

The proof of this convolution theorem is worth developing since the theorem itself is so useful. From equations 5.42 and 5.54

$$L(s) = \int_0^\infty f(t) \, e^{-st} dt$$

$$= \int_0^\infty \left[\int_{-\infty}^\infty f_1(\tau) f_2(t - \tau) \, d\tau \right] e^{-st} dt \ .$$

Changing the order of this integration

$$L(s) = \int_{-\infty}^\infty f_1(\tau) \left(\int_0^\infty f_2(t - \tau) \, e^{-st} \, dt \right) d\tau \ .$$

Using the time shifting theorem of equation 5.48

$$L(s) = \int_{-\infty}^\infty f_1(\tau) \, [e^{-s\tau} L_2(s)] \, d\tau$$

$$= L_2(s) \int_{-\infty}^\infty f_1(\tau) \, e^{-s\tau} d\tau$$

$$= L_1(s) \, L_2(s)$$

for $\tau \geq 0$, that is, for causal functions.

Some of the ways in which Laplace transform methods may be used to calculate and predict the responses of mechanical vibratory systems to a variety of driving functions will be discussed later in section 5.7. Firstly, a number of basic ideas about the damped harmonic oscillator need to be introduced.

Example 5.3

Obtain the Laplace transform of the unit area rectangular pulse centred at $t = t_0$ and of duration τ.

Solution

Here

$$f(t) = \frac{1}{\tau}\left\{U\left[t - \left(t_0 - \frac{\tau}{2}\right)\right] - U\left[t - \left(t_0 + \frac{\tau}{2}\right)\right]\right\}.$$

From equation 5.48

$$L(s) = \frac{1}{\tau}\left[\frac{e^{-s(t_0-\tau/2)}}{s} - \frac{e^{-s(t_0-\tau/2)}}{s}\right]$$

$$= \frac{e^{-st_0}}{s\tau}(e^{s\tau/2} - e^{-s\tau/2})$$

$$= \frac{e^{-st_0}\sinh(s\tau/2)}{(s\tau/2)}.$$

Note that in the limit of $\tau \to 0$ this expression approaches e^{-st_0}, the Laplace transform of a delta function at $t = t_0$, $\delta(t_0)$.

5.6 Linearity and Superposition

Schematically one can consider the driven damped harmonic oscillator as a black box as in figure 5.8. As described in chapter 4, the black box has as input the driving force $f(t)$ and as output or response function the steady state displacement of the body x_s. Consider only purely sinusoidal forcing functions $f_1(t)$ and $f_2(t)$. $f_1(t)$ acting alone produces a response x_{s1} while $f_2(t)$ results in x_{s2}. Examination of equation 4.1 shows that a forcing function consisting of a linear combination of $f_1(t)$ and $f_2(t)$

$$f(t) = k_1 f_1(t) + k_2 f_2(t) \tag{5.58}$$

where k_1 and k_2 are arbitrary constants will result in a total response

$$x_s = k_1 x_{s1} + k_2 x_{s2}. \tag{5.59}$$

Thus the total response is the same linear combination of the individual responses as the total input force is of the individual component forces.

HARMONIC ANALYSIS IN DAMPED HARMONIC OSCILLATOR SYSTEMS

Figure 5.8 A schematic view of the driven damped harmonic oscillator, a black-box with an input $f(t)$ and an output x_s.

The system is said to display linearity at least in regard to pure harmonic input forces. Furthermore, the total response described by equation 5.59 is the superposition of the individual responses while the total input force of equation 5.58 is the superposition of the component forces. Equations 5.58 and 5.59 might have any number of such elemental pure harmonic forces and resultant elemental responses.

For instance, the driving force might be the Fourier series expansion of equation 5.2 or 5.13. The total response would then be a linear combination or superposition of the set of harmonic responses — in effect a Fourier series expansion of x_s. If the input Fourier series is truncated then the response series will also be truncated and hence only an approximation to the actual response.

The ideas of linearity and superposition can be extended to any kind of periodic forcing function. Since periodic shock forces can be represented by a set of harmonics, the response to such an input can be calculated from the superposition of the array of spectral components.

In the case of driving force functions containing continuous spectra, such as isolated shocks, transform techniques are most powerful. They are basically the application of superposition over the frequency continuum, by integration, rather than the simple summation of the discrete harmonic responses discussed above.

5.7 Application of Laplace Transform Technique to a Linear System

As an example of the use of Laplace transform techniques consider the underdamped vibratory system of figure 4.1 to be subjected to a unit rectangular shock force of amplitude $1/\tau$ and duration τ, starting at $t = 0$. The solution or response may be divided into two parts, one for $0 \leq t \leq \tau$ and one for $t \geq \tau$.

The force equation for the first part may be written

$$M\ddot{x} + b\dot{x} + Cx = \frac{1}{\tau}U(t) \qquad 0 \leq t \leq \tau \qquad (5.60)$$

while the corresponding equation for the second part is

$$M\ddot{x} + b\dot{x} + Cx = \frac{1}{\tau}[U(t) - U(t - \tau)] \ . \qquad t \geq \tau \ . \quad (5.61)$$

Equation 5.60 may be solved using Laplace transforms. Take Laplace transforms of each side of the equation.

$$M[s^2 L(s) - sx(0) - \dot{x}(0)] + b[sL(s) - x(0)] + CL(s) = \frac{1}{\tau s} \ . \quad (5.62)$$

Taking the initial conditions for displacement $x(0) = 0$ and for velocity $\dot{x}(0) = 0$ and regrouping the terms in this equation produces

$$(Ms^2 + bs + C) L(s) = \frac{1}{\tau s} \ . \quad (5.63)$$

Dividing each term by M

$$(s^2 + 2\alpha s + \omega_0^2) L(s) = \left(\frac{1}{C_\tau}\right) \frac{\omega_0^2}{s} \ . \quad (5.64)$$

Therefore the Laplace transform of the displacement response is

$$L(s) = \frac{(1/C_\tau) \omega_0^2}{s[(s+\alpha)^2 + \omega_f^2]}$$

$$= \frac{(1/C_\tau)}{s} - \frac{(1/C_\tau) (s+\alpha)}{(s+\alpha)^2 + \omega_f^2} - \frac{(1/C_\tau) (\alpha/\omega_f)\omega_f}{(s+\alpha)^2 + \omega_f^2} \quad (5.65)$$

by taking partial fractions of the right-hand side of the equation. Using table 5.1, to take the inverse transform of this latter equation produces

$$x = \frac{1}{C_\tau}\left(1 - e^{-\alpha t} \cos \omega_f t - \left(\frac{\alpha}{\omega_f}\right) e^{-\alpha t} \sin \omega_f t\right) \quad (5.66)$$

which may be simplified to

$$x = \frac{1}{C_\tau}\left[1 - \frac{\omega_0}{\omega_f} e^{-\alpha t} \cos(\omega_f t - \gamma)\right] \quad (5.67)$$

where

$$\gamma = \tan^{-1}\left(\frac{\alpha}{\omega_f}\right) \ .$$

The second part of the solution may also be found using Laplace transforms. It will clearly be the superposition of the above solution and the solution to

HARMONIC ANALYSIS IN DAMPED HARMONIC OSCILLATOR SYSTEMS

$$M\ddot{x} + b\dot{x} + Cx = -\frac{1}{\tau}U(t-\tau) \ . \qquad t \geq \tau \ . \tag{5.68}$$

The solution to this latter equation may be written down by analogy with equation 5.67 thus

$$x = -\frac{1}{C\tau}\left\{1 - \frac{\omega_0}{\omega_f} e^{-\alpha(t-\tau)} \cos[\omega_f(t-\tau) - \gamma]\right\} \ . \tag{5.69}$$

The complete solution to the second part of the problem expressed in equation 5.61 will then be

$$x = \frac{\omega_0}{C\tau\omega_f} e^{-\alpha t}\left\{e^{-\alpha\tau} \cos[\omega_f(t-\tau) - \gamma] - \cos(\omega_f t - \gamma)\right\} .\tag{5.70}$$

But

$$e^{-\alpha\tau} = 1 - \alpha\tau - \frac{(\alpha\tau)^2}{2!} - \ldots$$

and if the system is slightly damped so that α is small or if the pulse is very short so that τ is very small then $e^{-\alpha\tau} = 1$.

Equation 5.70 then simplifies to

$$x = \left[\left(\frac{2\omega_0}{C\tau\omega_f}\right)\sin(\omega_f\tau/2)\right] e^{-\alpha t} \sin[\omega_f(t-\tau/2) - \gamma] \tag{5.71}$$

Thus after the pulse the response is an exponentially decaying sinusoid of frequency equal to the damped natural frequency of the system. The initial amplitude, however, depends in a complex way on the pulse duration τ. Note that if

$$\frac{\omega_f\tau}{2} = \pi, 2\pi, \ldots n\pi \qquad \text{for } n = 1, 2, 3, \ldots$$

or

$$\tau = \frac{2\pi}{\omega_f} \ldots \frac{2n\pi}{\omega_f}$$

then $\sin(\omega_f\tau/2) = 0$ and the amplitude is zero. Then what happens is n half cycles of free damped vibrations during the pulse with the mass at zero displacement and zero velocity at the end of the pulse. Thereafter, for $t > \tau$, the system remains at rest with no free vibrations.

5.8 Impulse Response and System Transfer Function

Consider the case of an isolated pulse driving force defined in the following way

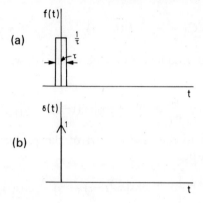

Figure 5.9 (a) An isolated pulse of duration τ and unit area. (b) a Dirac delta function or impulse which is an isolated pulse of infinitely narrow duration and unit area.

$$\begin{aligned} f(t) &= 0 & |t| &> \frac{\tau}{2} \\ &= \lim_{\tau \to 0} \left(\frac{1}{\tau}\right) & |t| &< \frac{\tau}{2} \\ &= \delta(t) \; . \end{aligned} \qquad (5.72)$$

This is the Dirac delta function, or an impulse at $t = 0$, shown in figure 5.9. An alternative definition of this function is through the integral

$$\int_{-\infty}^{\infty} \delta(t) \, dt = 1 \; . \qquad (5.73)$$

Clearly, the pulse $f(t)$ of figure 5.9(a) in the limit of $\tau \to 0$ becomes

$$f(t) \underset{\substack{\tau \to 0 \\ \text{(area constant)}}}{} = \delta(t) \qquad (5.74)$$

and

$$\int_{-\infty \text{ (pulse)}}^{\infty} f(t) \, dt = 1 \; . \qquad (5.75)$$

The Laplace transform of the Dirac delta function is thus

$$\begin{aligned} \mathscr{L} \delta(t) &= \int_{-\infty}^{\infty} \delta(t) \, e^{-st} \, dt \\ &= 1 \end{aligned} \qquad (5.76)$$

If a Dirac delta function driving force is applied to the damped harmonic oscillator of figure 4.1, the over-all force equation may be written

$$M\ddot{x} + b\dot{x} + Cx = \delta(t) \ . \tag{5.77}$$

Then

$$\ddot{x} + 2\alpha\dot{x} + \omega_0^2 x = \frac{\delta(t)}{M} \ . \tag{5.78}$$

Taking Laplace transforms of each side of this equation yields

$$(s^2 + 2\alpha s + \omega_0^2)\, X(s) = \frac{1}{M} \tag{5.79}$$

where

$$X(s) = \mathscr{L}(x) \ .$$

So

$$X(s) = \frac{1/M}{s^2 + 2\alpha s + \omega_0^2}$$

$$= \frac{1/M}{(s + \alpha)^2 + \omega_f^2} \tag{5.80}$$

using equation 3.20.

Figure 5.10 The system transfer function $H(s)$ multiplies the Laplace transform of the input forcing function $F(s)$, to yield the Laplace transform of the output function $X(s)$.

This particular Laplace transform, of the response of the system to the delta function driving force, is called the system transfer function, $H(s)$, illustrated in figure 5.10. More generally, the system transfer function is defined as the ratio of the transform of the system response to the transform of the driving force thus

$$H(s) = \frac{X(s)}{F(s)} \tag{5.81}$$

where $F(s)$ is the Laplace transform of the driving force. When as above $F(s) = 1$, then $H(s) = X(s)$.

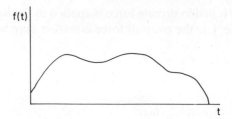

Figure 5.11 A general non-periodic forcing function $f(t)$.

The system response to the delta function, $h(t)$, is the inverse transform of the system transfer function

$$h(t) = \mathscr{L}^{-1}[H(s)] \ . \tag{5.82}$$

This is the impulse response of the system.

Knowledge of the impulse response of a system is most useful in determining the response of the system to any general driving force function such as that illustrated in figure 5.11. Let $F(s)$ be the Laplace transform of this function $f(t)$. The response $x(t)$ is sought. From equation 5.81 the transform of this response is

$$X(s) = H(s) \, F(s) \ . \tag{5.83}$$

It was shown in equations 5.54 and 5.57 that the inverse transform of $X(s)$, $x(t)$ is the convolution of the impulse response $h(t)$ and the driving force function $f(t)$ thus

$$\begin{aligned} x(t) &= \mathscr{L}^{-1}[X(s)] \\ &= \int_{-\infty}^{\infty} h(\tau) \, f(t - \tau) \, \mathrm{d}\tau \\ &= \int_{-\infty}^{\infty} f(\tau) \, h(t - \tau) \, \mathrm{d}\tau \ . \end{aligned} \tag{5.84}$$

Thus the response of the system may be obtained by one or other of these straightforward integrations. Even if the functions $h(t)$ and $f(t)$ are not simple, the use of digital computer methods of integration renders this convolution method a very powerful and widely applicable way of determining the behaviour of a linear system.

Example 5.4

Use the convolution method to obtain the system response to the rectangular shock force of equation 5.61.

Solution

From equation 5.80 the system transfer function $H(s)$ is

$$H(s) = \frac{1/M}{(s + \alpha)^2 + \omega_f^2} \; .$$

Using table 5.1, the impulse response of the system $h(t)$ is found

$$h(t) = \frac{1}{M\omega_f} e^{-\alpha t} \sin \omega_f t \; .$$

The input forcing function in this instance is

$$f(t) = \frac{1}{\tau}[U(t) - U(t - \tau)] \; . \qquad \text{for } t \geq \tau$$

By the convolution theorem of equation 5.84, the response of the system $x(t)$ is

$$x(t) = \frac{1}{\tau} \int_{-\infty}^{\infty} h(y) \, [U(t - y) - U(t - \tau - y)] \, dy$$

where y is used as the running variable to avoid confusion with the shock duration.

$$x(t) = \frac{1}{M\tau\omega_f} \left(\int_0^t e^{-\alpha y} \sin \omega_f y \, dy - \int_0^{t-\tau} e^{-\alpha y} \sin \omega_f y \, dy \right)$$

$$= \frac{1}{M\tau\omega_f\omega_0^2} [e^{-\alpha y}(-\alpha \sin \omega_f y - \omega_f \cos \omega_f y)] \Big|_0^t$$

$$-e^{-\alpha y}(-\alpha \sin \omega_f y - \omega_f \cos \omega_f y) \Big|_0^{t-\tau}]$$

$$= \frac{\omega_0}{C\tau\omega_f} \{-e^{-\alpha t} \cos(\omega_f t - \gamma) + e^{-\alpha(t-\tau)} \cos[\omega_f(t-\tau) - \gamma]\}$$

which is identical with equation 5.70.

5.9 Experiments

(a) Fourier Analysis with a Frequency Analyser

A frequency analyser is an electronic apparatus equipped with a band-pass filter the centre-frequency of which can be continuously varied, followed by a measuring meter. The meter reading is the voltage level of the components in the input signal at the centre-frequency which is manually set.

Various types of periodic voltages are available from simple signal generators, sinusoidal, square, triangular, saw-tooth, rectangular, and all of these may be analysed by the analyser into the Fourier components. The repetitive rectangular signal allows the spectrum of a similarly shaped isolated pulse to be deduced.

A repeated sound shock can be transduced with a microphone and the voltage signal analysed. A vibration shock, transduced with an accelerometer, may likewise be analysed, if it is repeated.

(b) Fourier Analysis with a Spectrum Analyser

A spectrum analyser electronically scans through the frequency range and presents the line spectrum of a periodic signal as a display on an oscilloscope. It is virtually instantaneous and does not require repetitions of a pulsed signal to yield the relevant continuous spectrum.

Problems

5.1 Obtain the first four terms in the Fourier series representation of the following functions

(a) $f(t) = \cos t \quad -\dfrac{\pi}{2} \leq t \leq \dfrac{\pi}{2}$

$ = 0 \quad -\pi \leq t \leq -\dfrac{\pi}{2}$ and $\dfrac{\pi}{2} \leq t \leq \pi$

a half-way rectified sine wave;

(b) $f(t) = |\cos 10t|$

a full-wave rectified sine wave;

(c) $f(t) = 1 \quad 0 \leq t \leq \dfrac{T}{2}$

$ = -1, \quad \dfrac{T}{2} \leq t \leq T$

a square wave;

(d) $f(t) = kt \quad -\dfrac{T}{2} \leq t \leq +\dfrac{T}{2}$

a saw-tooth triangular wave;

(e) $f(t) = 0 \quad -\dfrac{T}{2} \leq t \leq -\dfrac{\tau}{2}$

$\quad = \cos \omega_c t \quad -\dfrac{\tau}{2} \leq t \leq \dfrac{\tau}{2}$

$\quad = 0 \quad \dfrac{\tau}{2} \leq t \leq \dfrac{T}{2}$

a pulse modulated sine wave.

5.2 Find the Fourier transform for an isolated cycle of each of the functions of problems 5.1.

5.3 Find the inverse Laplace transforms of the following functions by first decomposing them into partial fractions:

(a) $\dfrac{a}{s(s + a)}$

(b) $\dfrac{s^2 + 2}{s^3}$

(c) $\dfrac{2s^2 + a^2}{s(s^2 + a^2)}$

(d) $\dfrac{a^2}{s(s^2 + a^2)}$

(e) $\dfrac{a^3}{s(s + a)^3}$

5.4 A horizontally moving damped mechanical system consists of a mass of 9 kg, a spring of constant 20 N/m and a damper of coefficient 12 kg/s. Find the system transfer function. Take the system to be initially at rest at the equilibrium position. The system is subjected to a half-wave rectified sine wave thus

$f(t) = 0 \quad -0.5 \leq t \leq -0.25$

$\quad = \cos 2\pi t \quad -0.25 \leq t \leq 0.25$

$\quad = 0 \quad 0.25 \leq t \leq 0.5$

Find the response of the system by convolution.

5.5 A spring of constant 200 N/m hangs vertically. A mass of 10 kg is attached to the bottom end of the spring and initially is held at the position where the spring is not stretched. At $t = 0$ the mass is released from rest. Use Laplace transforms to derive the subsequent displacement of the mass.

5.6 Get the response of the system with the following system transfer function

$$H(s) = \frac{1}{s + 1}$$

for each of the following causal forcing functions:

(a) $f(t) = \delta(t)$
(b) $f(t) = U(t)$
(c) $f(t) = \cos \omega t$ (ω constant)
(d) $f(t) = kt$ (k constant).

5.7 Find the response of the system with system transfer function

$$H(s) = \frac{s}{(s + 1)^2}$$

to each of the driving force functions of problem 5.6.

5.8 For each of the driving force functions in problem 5.6 find the response of the system with transfer function

$$H(s) = \frac{s}{(s + 1)(s + 2)}.$$

Further Reading

Bishop, R.E.D., *Vibration* (Cambridge University Press, Cambridge, 1965)
Braddick, H.J.J., *Vibrations, Waves and Diffraction* (McGraw-Hill, London, 1965)
Church, A.H., *Mechanical Vibrations,* 2nd edition (Wiley, New York, 1963)
Dimarogonas, A.D., *Vibration Engineering* (West, St. Paul, 1976)
Hobbie, R.K., *Intermediate Physics for Medicine and Biology* (Wiley, New York, 1978)
Main, I.G., *Vibrations and Waves in Physics* (Cambridge University Press, Cambridge, 1978)
Papoulis, A., *Signal Analysis* (McGraw-Hill, New York, 1977)
Papoulis, A., *The Fourier Integral and its Applications* (McGraw-Hill, New York, 1962)
Steidel, R.F. Jr., *An Introduction to Mechanical Vibrations,* 2nd edition (Wiley, New York, 1979)
Williams, J., *Laplace Transforms* (Allen & Unwin, London, 1973)

6 Coupled Harmonic Oscillators

6.1 Introduction and Objectives

The aim in this chapter is to discuss the situation in which two mechanical oscillators interact with each other and energy can be interchanged between them.

After reading this chapter the student should be able to

(a) draw up the system equations for a pair of mechanical oscillators coupled via a spring member;
(b) define the modes of such a coupled scheme;
(c) describe the displacements of the two masses in terms of the interference pattern of two harmonics;
(d) describe the energies in the over-all system in terms of the mode coordinates;
(e) describe the resonances in the modes of the syste;
(f) draw up electrical analogue circuits to simulate the coupled mechanical system;

6.2 Coupled Oscillating Systems — Modes

The system depicted in figure 6.1 consists of two identical undamped oscillators each of mass M, spring of constant C and damping coefficient b, coupled by means of a third spring of constant c. Let x_1 be the

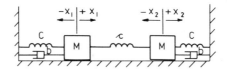

Figure 6.1 Two identical damped harmonic oscillators coupled by a spring of constant c joined at each end to one of the masses.

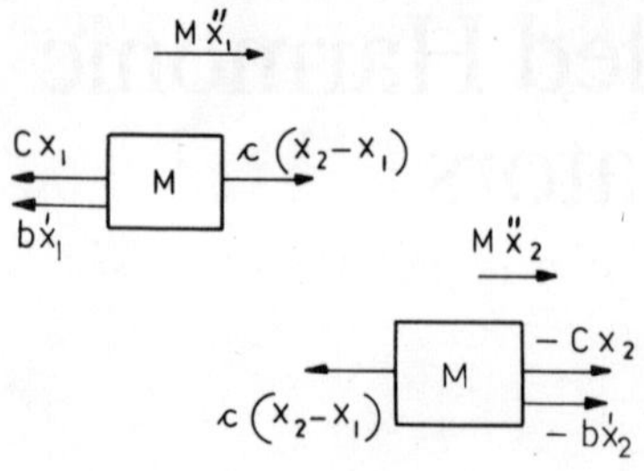

Figure 6.2 The free-body diagrams for the two masses in the spring-coupled harmonic vibrators. Each mass experiences spring and damper forces which result in acceleration.

along the horizontal direction from its equilibrium position, and x_2 that of the right-hand body. The resulting free-body diagrams are shown in figure 6.2. For the left-hand body the force relation is

$$c(x_2 - x_1) - Cx_1 - bx_1' = M x_1'' \tag{6.1}$$

and the equation of motion is therefore

$$x_1'' + \frac{b}{M} x_1' + \left(\frac{C + c}{M}\right) x_1 - \frac{c}{M} x_2 = 0 . \tag{6.2}$$

The two displacements x_1 and x_2 are thus interlinked. Similarly for the right-hand body, the force relationship may be written down from the free-body diagram of figure 6.2

$$-Cx_2 - c(x_2 - x_1) - bx_2' = M x_2'' \tag{6.3}$$

from which the equation of motion emerges by dividing each term by M

$$x_2'' + \frac{b}{M} x_2' + \left(\frac{C + c}{M}\right) x_2 - \frac{c}{M} x_1 = 0 . \tag{6.4}$$

Once more there is coupling between the two displacements.
Adding each side of equations 6.2 and 6.4 yields

$$x_1'' + x_2'' + \frac{b}{M}(x_1' + x_2') + \frac{C}{M}(x_1 + x_2) = 0 \tag{6.5}$$

or

$$p_1'' + \frac{b}{M} p_1' + \frac{C}{M} p_1 = 0 \tag{6.6}$$

COUPLED HARMONIC OSCILLATORS

where

$$p_1 = x_1 + x_2 . \tag{6.7}$$

Subtracting each side of equation 6.4 from each side of equation 6.2 gives

$$\ddot{x}_1 - \ddot{x}_2 + \frac{b}{M}(\dot{x} - \dot{x}_2) + \left(\frac{C + 2c}{M}\right)(x_1 - x_2) = 0 \tag{6.8}$$

or

$$\ddot{p}_2 + \frac{b}{M}\dot{p}_2 + \left(\frac{C + 2c}{M}\right)p_2 = 0 \tag{6.9}$$

where

$$p_2 = x_1 - x_2 . \tag{6.10}$$

Referring back to equation 3.17, it is apparent that equations 6.6 and 6.9 are the equations of damped harmonic oscillations. p_1 and p_2 are not here physical displacements of either of the masses in the system, rather are they the instantaneous sum and difference respectively of the two physical displacements of the two masses. They are the 'displacements' of the two modes of the system. Equations 6.6 and 6.9 are the equations of motion of the two modes. Solutions to equations 6.6 and 6.9 take the following forms after equation 3.18

$$p_1 = P_1 e^{-\alpha t} \cos(\omega_{01} t + \gamma_1) \tag{6.11}$$

and

$$p_2 = P_2 e^{-\alpha t} \cos(\omega_{02} t + \gamma_2) . \tag{6.12}$$

P_1 and P_2 are amplitudes while γ_1 and γ_2 are phase angles set by the initial conditions of the system. The parameters of the system itself determine α, ω_{01} and ω_{02} thus

$$\alpha = \frac{b}{2M} \tag{6.13}$$

$$\omega_{01} = \sqrt{\left(\frac{C}{M}\right)} \tag{6.14}$$

and

$$\omega_{02} = \sqrt{\left(\frac{C + 2c}{M}\right)} . \tag{6.15}$$

Note that from equations 6.7 and 6.10 the physical displacements of the two bodies x_1 and x_2 may be derived from the mode 'displacements' p_1 and p_2 thus

$$x_1 = \tfrac{1}{2}(p_1 + p_2) \tag{6.16}$$

and

$$x_2 = \tfrac{1}{2}(p_1 - p_2) . \tag{6.17}$$

A number of cases of interest can arise in real coupled mechanical systems.

(a) c negligibly small

If $c \ll C$, then the two basic equations of motion, 6.2 and 6.4, reduce to two identical but independent uncoupled harmonic oscillations. This is the case of flexible to negligible coupling.

(b) c very large

If $c \gg C$, then $x_1 = x_2$ and the system reduces to that depicted in figure 2.9. The motion of each mass is identically coupled to that of the other. This is the case of rigid coupling.

(c) If the two bodies are given an equal initial displacement A_0, to the right and then released from rest

Now at $t = 0$, $p_1(0) = 2A_0$ and $p_2(0) = 0$. Also $\dot{p}_1(0) = 0$ and $\dot{p}_2(0) = 0$. Therefore from equations 6.11 and 6.12

$$2A_0 = P_1 \cos \gamma_1$$

$$0 = -\omega_{01} P_1 \sin \gamma_1 - \alpha P_1 \cos \gamma_1 .$$

Therefore

$$\gamma_1 = \tan^{-1}\left(-\frac{\alpha}{\omega_{01}}\right) \tag{6.18}$$

and

$$P_1 = \frac{2A_0}{\omega_{01}} \sqrt{\left(\alpha^2 + \omega_{01}^2\right)} . \tag{6.19}$$

Also it can be shown that

COUPLED HARMONIC OSCILLATORS

$$P_2 = 0 \text{ and } \gamma_2 = 0$$

Under these circumstances mode 1 only is excited while mode 2 is at rest. This means that the two masses move in phase with each other and with equal displacement, that is

$$x_1 = x_2$$

$$= \tfrac{1}{2} p_1$$

$$= \frac{A_0}{\omega_{01}} \sqrt{\left(\alpha^2 + \omega_{01}^2\right)} e^{-\alpha t} \cos\left[\omega_{01} t + \tan^{-1}\left(-\frac{\alpha}{\omega_{01}}\right)\right]. \quad (6.20)$$

The coupling spring constant c thus plays no role in determining the characteristics of the motion regardless of its value.

(d) If the two bodies are displaced equal displacements A_0 in opposite directions and then released from rest

Here at $t = 0$, $x_1(0) = A_0$ and $x_2(0) = -A_0$ so that $p_1(0) = 0$, $p_2(0) = 2A_0$, $\dot{p}_1(0) = 0$ and $\dot{p}_2(0) = 0$. By substituting in equations 6.11 and 6.12 it can be shown that

$$P_1 = 0 \text{ and } \gamma_1 = 0 \text{ while}$$

$$P_2 = \frac{2A_0}{\omega_{02}} \sqrt{\left(\alpha^2 + \omega_{02}^2\right)} \quad (6.21)$$

and

$$\gamma_2 = \tan^{-1}\left(-\frac{\alpha}{\omega_{02}}\right). \quad (6.22)$$

In this case only mode 2 is excited and

$$x_1 = -x_2$$

$$= \tfrac{1}{2} p_2$$

$$= \frac{A_0}{\omega_{02}} \sqrt{\left(\alpha^2 + \omega_{02}^2\right)} e^{-\alpha t} \cos\left[\omega_{02} t + \tan^{-1}\left(-\frac{\alpha}{\omega_{02}}\right)\right] \quad (6.23)$$

The two bodies have equal displacement and frequency but move in antiphase (π rad out of phase) with each other.

(e) If the right-hand body is displaced by $\sqrt{(2)}A_0$ and the left-hand body by zero and then both are released from rest

Here at $t = 0$, $x_1(0) = 0$, $x_2(0) = \sqrt{(2)}A_0$ so that $p_1(0) = \sqrt{(2)}A_0$ $p_2(0) = -\sqrt{(2)}A_0$. Also $\dot{p}_1(0) = 0$ and $\dot{p}_2(0) = 0$. Substituting in equations 6.11 and 6.12 produces

$$P_1 = \frac{\sqrt{(2)}A_0}{\omega_{01}} \sqrt{(\alpha^2 + \omega_{01}^2)} \tag{6.24}$$

$$\gamma_1 = \tan^{-1}\left(-\frac{\alpha}{\omega_{01}}\right) \tag{6.25}$$

$$P_2 = -\frac{\sqrt{(2)}A_0}{\omega_{02}} \sqrt{(\alpha^2 + \omega_{02}^2)} \tag{6.26}$$

and

$$\gamma_2 = \tan^{-1}\left(-\frac{\alpha}{\omega_{02}}\right). \tag{6.27}$$

In this case

$$x_1 = \frac{A_0}{\sqrt{(2)}} e^{-\alpha t} \left[\frac{\sqrt{(\alpha^2 + \omega_{01}^2)}}{\omega_{01}} \cos(\omega_{01}t + \gamma_1) \right.$$

$$\left. - \frac{\sqrt{(\alpha^2 + \omega_{02}^2)}}{\omega_{02}} \cos(\omega_{02}t + \gamma_2) \right]. \tag{6.28}$$

and

$$x_2 = \mp \frac{A_0}{\sqrt{(2)}} e^{-\alpha t} \left[\frac{\sqrt{\alpha^2 + \omega_{01}^2}}{\omega_{01}} \cos(\omega_{01}t + \gamma_1) \right.$$

$$\left. + \frac{\sqrt{(\alpha^2 + \omega_{02}^2)}}{\omega_{02}} \cos(\omega_{02}t + \gamma_2) \right]. \tag{6.29}$$

To obtain an insight into this coupled oscillatory motion, consider the simplified situation of negligible damping ($\alpha = 0$) and very weak coupling (c small). Then $\gamma_1 = 0 = \gamma_2$ and ω_{01} is almost equal to ω_{02}. Then

$$x_1 = \frac{A_0}{\sqrt{(2)}\omega_{01}} (\cos \omega_{01}t - \cos \omega_{02}t)$$

$$= -\frac{\sqrt{(2)}A_0}{\omega_{01}} \sin\left(\frac{\omega_{02} + \omega_{01}}{2}\right)t \sin\left(\frac{\omega_{02} - \omega_{01}}{2}\right)t \tag{6.30}$$

and

$$x_2 = \frac{A_0}{\sqrt{2}\,\omega_{02}} (\cos \omega_{01} t + \cos \omega_{02} t)$$

$$= \frac{\sqrt{2}\,A_0}{\omega_{02}} \cos\left(\frac{\omega_{02} + \omega_{01}}{2}\right) t \cos\left(\frac{\omega_{02} - \omega_{01}}{2}\right) t \ . \tag{6.31}$$

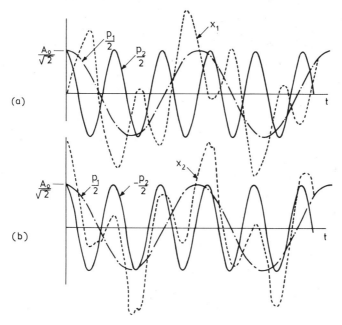

Figure 6.3 The displacement of the left-hand mass x_1 in free vibration may be calculated as in (a) by adding together at each instant one half of each of the harmonic mode displacements p_1 and p_2. In (b) the displacement of the right-hand mass x_2 may be calculated by subtracting one half of p_2 from one half of p_1 at each instant.

Figure 6.3 illustrates these two motions of the bodies. Broadly the two displacements are in quadrature to each other ($\pi/2$ rad out of phase). First one mass experiences large amplitude of displacement while the other has a low amplitude. Then the second mass acquires a high amplitude while the amplitude of the first declines. There is a constant energy exchange between the two oscillators with each one serving in turn as a source or driver with the other acting as an energy sink or driven system. The weaker the coupling and hence the smaller is ($\omega_{02} - \omega_{01}$), the more cycles of high amplitude and high energy each separate oscillator has before the other takes over in this role.

Example 6.1

In a coupled mechanical system such as that of figure 6.1, each mass is 1 kg and each main spring has a constant of 80 N/m. The coupling spring constant is to be determined. Damping is negligible. The following observations are made on the system. At $t = 0$, $x_1 = 0.05$ m, $x_2 = 0.01$ m, $\dot{x}_1 = 0 = \dot{x}_2$. At $t = 0.15$ s, $x_1 = 0$.

Solution

From equation 6.14, $\omega_{01} = \sqrt{(C/M)} = 10$ rad/s. By substituting in equations 6.11 and 6.12, $\gamma_1 = 0 = \gamma_2$ and $P_1 = 0.06$ m while $P_2 = 0.04$ m. So

$$x_1(t) = 0.03 \cos 10t + 0.02 \cos \omega_{02} t$$

and this $= 0$ at $t = 0.15$ s. So

$$0 = 0.03 \cos 1.5 + 0.02 \cos \omega_{02} t$$

$$\cos \omega_{02} t = -0.11$$

$$\omega_{02} = 11.2 \text{ rad/s} = \sqrt{\left(\frac{100 + 2c}{1}\right)} \text{ from equation 6.15}$$

and

$$c = 12.7 \text{ N/m} .$$

6.3 Energy in the Coupled System

The total energy in the system of figure 6.1 may be calculated by first finding the kinetic and potential energies. The kinetic energy E_k may be written

$$E_k = \tfrac{1}{2} M \dot{x}_1^2 + \tfrac{1}{2} M \dot{x}_2^2$$

$$= \tfrac{1}{2} M (\dot{x}_1^2 + \dot{x}_2^2). \tag{6.32}$$

COUPLED HARMONIC OSCILLATORS

The potential energy is given by

$$E_p = \tfrac{1}{2}Cx_1^2 + \tfrac{1}{2}Cx_2^2 + \tfrac{1}{2}c(x_2 - x_1)^2$$
$$= \tfrac{1}{2}(C + c)x_1^2 + \tfrac{1}{2}(C + c)x_2^2 - cx_1x_2 \ . \tag{6.33}$$

If the so-called mode co-ordinates or normal co-ordinates of the coupled system ψ_1 and ψ_2 are defined thus

$$\psi_1 = \sqrt{\left(\frac{M}{2}\right)} p_1 \tag{6.34}$$

and

$$\psi_2 = -\sqrt{\left(\frac{M}{2}\right)} p_2 \tag{6.35}$$

the two energy equations 6.32 and 6.33 may be rewritten. But first note that from equations 6.16 and 6.17

$$x_1 = \frac{1}{\sqrt{(2M)}}(\psi_1 - \psi_2) \tag{6.36}$$

and

$$x_2 = \frac{1}{\sqrt{(2M)}}(\psi_1 + \psi_2) \ . \tag{6.37}$$

Therefore

$$E_k = \tfrac{1}{2}\dot{\psi}_1^2 + \tfrac{1}{2}\dot{\psi}_2^2 \tag{6.38}$$

and

$$E_p = \tfrac{1}{2}\omega_{01}^2\psi_1^2 + \tfrac{1}{2}\omega_{02}^2\psi_2^2 \ . \tag{6.39}$$

The total energy in the system E_t is then

$$E_t = E_k + E_p$$
$$= \tfrac{1}{2}(\dot{\psi}_1^2 + \omega_{01}^2\psi_1^2) + \tfrac{1}{2}(\dot{\psi}_2^2 + \omega_{02}^2\psi_2^2) \tag{6.40}$$

the sum of the energies in the two modes of the system.

Thus the modes and mode co-ordinates of the coupled system, while conceptual and not attributable to discrete components in the system but to the system as a whole, play a useful role in describing the behaviour of the elements of the system.

Example 6.2

Find the total energy in the coupled system of example 6.1.

Solution

From equations 6.34 and 6.35

$$\psi_1 = \sqrt{\left(\frac{M}{2}\right)} p_1 = \frac{0.06}{\sqrt{2}} \cos 10t$$

$$\psi_2 = -\sqrt{\left(\frac{M}{2}\right)} p_2 = -\frac{0.04}{\sqrt{2}} \cos 11.2t$$

So from equation 6.40 the total energy in the system is

$$E_t = \tfrac{1}{2}(\dot{\psi}_1^2 + \omega_{01}^2 \psi_1^2) + \tfrac{1}{2}(\dot{\psi}_2^2 + \omega_{02}^2 \psi_2^2)$$

$$= 0.18 + 0.05$$

$$= 0.23 \text{ J} \;.$$

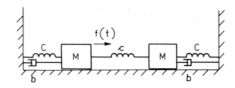

Figure 6.4 Spring-coupled damped harmonic oscillators with one of the masses driven by an external force $f(t)$.

6.4 Resonances in the Coupled System

If the left-hand body of figure 6.4 is subjected to a sinusoidal driving force $f(t) = F_0 \cos \omega t$, the force equations of the two masses may be written down from the free-body diagrams of figure 6.5

$$F_0 \cos\omega t - b\dot{x}_1 - (C + c)x_1 + cx_2 = M\ddot{x}_1 \tag{6.41}$$

and

$$-b\dot{x}_2 - (C + c)x_2 + cx_1 = M\ddot{x}_2 \tag{6.42}$$

COUPLED HARMONIC OSCILLATORS

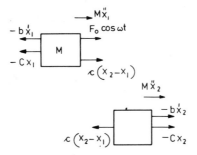

Figure 6.5 The free-body diagram for the two masses in the spring-coupled system with one mass subjected to a single-frequency alternating force.

Substituting for x_1 and x_2 from equations 6.36 and 6.37 and in turn adding each term in these two equations and subtracting them, the equations of 'motion' of the mode co-ordinates ψ_1 and ψ_2 emerge

$$\ddot{\psi}_1 + 2\alpha\dot{\psi}_1 + \omega_{01}^2\psi_1 = \frac{F_0}{\sqrt{(2M)}}\cos\omega t \qquad (6.43)$$

and

$$\ddot{\psi}_2 + 2\alpha\dot{\psi}_2 + \omega_{02}^2\psi_2 = -\frac{F_0}{\sqrt{(2M)}}\cos\omega t$$

$$= \frac{F_0}{\sqrt{(2M)}}\cos(\omega t - \pi) \ . \qquad (6.44)$$

In the steady state, therefore, the two mode co-ordinates will vary sinusoidally at the driving frequency. Following equation 4.8

$$\psi_1 = \Psi_{1s}\cos(\omega t - \gamma_{1s})$$

and $\qquad\qquad\qquad\qquad\qquad\qquad\qquad\qquad\qquad\qquad\qquad\qquad$ (6.45)

$$\psi_2 = \Psi_{2s}\cos(\omega_t - \gamma_{2s}) \ .$$

The amplitude terms Ψ_{1s} and Ψ_{2s} are given from equation 4.9

$$\Psi_{1s} = \frac{\sqrt{(M/2)}F_0}{\sqrt{[(\omega_{01}^2 - \omega^2)^2 + (2\alpha\omega)^2]}} \qquad (6.46)$$

and

$$\Psi_{2s} = \frac{\sqrt{(M/2)}F_0}{\sqrt{[(\omega_{02}^2 - \omega^2)^2 + (2\alpha\omega)^2]}} \ . \qquad (6.47)$$

The phase angles γ_{1s} and γ_{2s} may be written down with reference to equation 4.10

$$\gamma_{1s} = \tan^{-1}\left(\frac{2\alpha\omega}{\omega_{01}^2 - \omega^2}\right) \tag{6.48}$$

and

$$\gamma_{2s} = -\pi + \tan^{-1}\left(\frac{2\alpha\omega}{\omega_{02}^2 - \omega^2}\right). \tag{6.49}$$

Thus each mode resonates when the driving frequency is in the vicinity of the undamped natural frequency of that mode. The mode amplitudes Ψ_{1s} and Ψ_{2s} and the phase angles γ_{1s} and γ_{2s} vary with driving frequency as shown in figure 6.6.

The actual steady-state motions of the two bodies may be calculated from the steady-state mode co-ordinates of equations 6.45 and 6.46 using

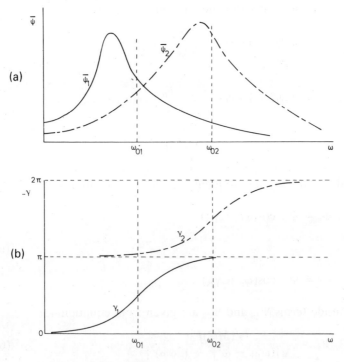

Figure 6.6 (a) The variation of the amplitudes of the displacements of the mode co-ordinates with frequency of the driving force. (b) The dependence on frequency of the phase angles of the displacements of the mode co-ordinates.

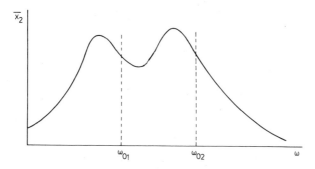

Figure 6.7 The amplitude of the displacement of the right-hand mass in the spring-coupled system as a function of driving frequency.

equations 6.36 and 6.37. For instance, the amplitude of x_2, the displacement of the right-hand mass, varies along the lines indicated in figure 6.7.

6.5 Electrical Analogues

Coupled mechanical oscillatory systems may be modelled quite conveniently with multi-loop electrical circuits. An electrical element shared between two loops provides the coupling. The spring-coupled pair of mechanical vibrators of figure 6.1 may be shown to be directly simulated by the two-loop electrical circuit of figure 6.8. Since a spring of constant c

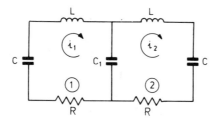

Figure 6.8 A double-loop electrical circuit in which the two loops share a capacitor is the exact analogue of the spring-coupled harmonic oscillators of figure 6.1.

provides the mechanical coupling, the electrical analogue of a spring, a capacitor of capacitance C_1 ($= 1/c$) is the electrical couplant.

The voltage relationships for each loop may be written thus:

Loop 1

$$L\ddot{q}_1 + R\dot{q}_1 + \left(\frac{1}{C} + \frac{1}{C_1}\right)q_1 - \frac{1}{C_1}q_2 = 0 \ . \tag{6.50}$$

Loop 2

$$L\ddot{q}_2 + R\dot{q}_2 + \left(\frac{1}{C} + \frac{1}{C_1}\right)q_2 - \frac{1}{C_1}q_1 = 0 \ . \tag{6.51}$$

Clearly these two equations are identical in form with equations 6.1 and 6.3 respectively if the customary analogies prescribed in table 2.1 are used.

The circuit of figure 6.8 can be readily assembled and used to study the behaviour of the coupled mechanical system. Introduction of a variable frequency signal generator into loop number 1 allows the resonant behaviour discussed in section 6.4 to be studied.

The coupled system described in this chapter consists of a pair of identical harmonic oscillators. This allows readily soluble mathematics while still conveying the basic aspects of the behaviour of coupled systems under different conditions. Use of an electrical analogue set-up allows one to depart from this simple case to the study of coupling between a variety of dissimilar oscillators.

Example 6.4

The two horizontally moving mechanical oscillators in figure 6.9 are coupled by the common damper of damping coefficient b. Write down the force equations for the two oscillators, find the mode co-ordinates of the coupled system and draw an electrical analogue circuit for it.

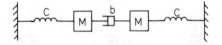

Figure 6.9 A system in which the two harmonic oscillators are coupled by a damping element of coefficient b.

Solution

For the left-hand mass

$$M\ddot{x}_1 + Cx_1 - b(\dot{x}_2 - \dot{x}_1) = 0 \quad \text{or}$$

$$M\ddot{x}_1 + b\dot{x}_1 + Cx_1 - b\dot{x}_2 = 0 \ .$$

COUPLED HARMONIC OSCILLATORS

For the right-hand mass

$$M\ddot{x}_2 + b\dot{x}_2 + Cx_2 - b\dot{x}_1 = 0.$$

Adding these two equations and dividing each term by M yields

$$\ddot{p} + \omega_{01}^2 p_1 = 0$$

where $p_1 = x_1 + x_2$ and $\omega_{01}^2 = C/M$. Similarly, by subtraction is found

$$\ddot{p}_2 + 4\alpha \dot{p}_2 + \omega_{01}^2 p_2 = 0$$

where $p_2 = x_1 - x_2$ and $\alpha = b/2M$. The mode co-ordinates are given from equations 6.34 and 6.35

$$\psi_1 = \sqrt{\left(\frac{M}{2}\right)} p_1 \text{ and } \psi_2 = -\sqrt{\left(\frac{M}{2}\right)} p_2.$$

From the force equations, and bearing in mind the usual analogies, the circuit of figure 6.10 is an analogue of the mechanical system. The symbols for the mechanical elements are written beside the electrical elements.

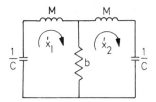

Figure 6.10 A two-loop electrical circuit in which the loops share a resistor is an analogue of the system of figure 6.9.

6.6 Experiments

(a) Spring-coupled Oscillators

Two identical oscillators of known mass and spring constant may be coupled as in figure 6.1 on a smooth table by means of a spring of constant c whether this is known or not. The types of coupling and initial conditions outlined in section 6.2 (a) to (e) may be investigated. If one of the bodies, say the left-hand one, is attached to the slider on a potential divider, a voltage proportional to x_1 can be generated. Depending on the initial conditions this should correspond to equation 6.20, 6.23 or 6.28. The mode frequencies and phase angles may be derived from such recorded voltages.

(b) Forced Spring-coupled Oscillators

A variable-frequency vibrator may be used to drive the left-hand mass in the coupled system as in figure 6.4. Potential divider displacement transducers may be used to record the displacements x_1 and x_2 of the two masses. The mode co-ordinates can be electronically derived from these and the mode resonances studied as in figure 6.6 by varying the driving frequency.

(c) Coupled Electrical Circuits

A two-loop electrical circuit such as that in figure 6.8 can be assembled with any desired component values. Initial conditions can be set by charging the capacitors to any required voltage levels. Subsequent behaviour of the system can be studied with an oscilloscope and A.C. ammeters and voltmeters.

A signal generator in one of the loops of the system allows the behaviour of the driven system to be studied.

An inductively coupled system is an analogue of a mass-coupled scheme and a resistance-coupled circuit is an analogue of a damper-coupled system. Either of these possibilities may be readily incorporated in an electrical circuit for study.

Problems

6.1 Two identical oscillators as in figure 6.1, each of mass 100 kg and spring of constant 500 N/m, are coupled by a spring of constant 50 N/m. Find the two natural mode frequencies. If at $t = 0$ the left-hand mass has a displacement of 0.1 m and a velocity of 1 m/s while the second mass is at rest and zero displacement, write down the subsequent mode 'displacements' and the actual displacements of the masses.

6.2 If in the above system the total energy must decline by 50 per cent in 10 s, find the value of the identical damping coefficients that must be introduced into the coupled oscillators.

6.3 An automobile consists of body/engine mass on suspension springs on a chassis mass on the spring of the tyres. Given the following data find the

COUPLED HARMONIC OSCILLATORS

two natural mode frequencies: body mass 1000 kg; chassis mass 300 kg; spring constant of suspension 60 kN/m; spring constant of tyres 700 kN/m.

6.4 Draw an electrical circuit to simulate the system of problem 6.3. If at $t = 0$ the body mass is displaced downwards by 0.01 m while the chassis mass is displaced downwards by 0.001 m, with both of them then at rest, find the expressions for the subsequent displacements of body and chassis.

6.5 The automobile in problem 6.3 is driven over a road with sinusoidal undulations (10 mm crest-to-crest). Find the speeds at which the mode resonances are stimulated.

6.6 A mass of 10 kg is suspended by two springs. One of these springs is of constant 100 N/m and its top end is rigidly fixed. The other spring has a constant 100 N/m and its top attaches to a second mass of 10 kg which in turn is suspended on the bottom of a second spring constant 100 N/m. What are the two natural mode frequencies of this system?

6.7 At $t = 0$, the lower mass in problem 6.6 has an upwards displacement of 1 cm and zero velocity while the other mass is at rest at its equilibrium location. Describe the subsequent motion of the two masses.

6.8 Draw an electrical circuit to simulate the system of problem 6.6 and find the damping coefficients to act in parallel with each spring so that either of the two modes is critically damped. In each case what will the damping ratio of the other mode be?

6.9 Three masses M_1, M_2 and M_1 are connected along the x-axis by two springs each of constant C. Calculate the two mode frequencies and draw an electrical circuit to simulate the system.

6.10 Two identical simple pendulums each with bob of mass M and stiff rod (hinged at the top) of length L are mounted close by each other. A spring of constant C is connected between the two rods a distance w down from the hinges. Show that the mode natural frequencies of this coupled system are

$$f_1 = \frac{1}{2\pi}\sqrt{\left(\frac{g}{L}\right)} \text{ and } f_2 = \frac{1}{2\pi}\sqrt{\left(\frac{g}{L} + 2\frac{w^2 C}{L^2 M}\right)}.$$

6.11 Calculate the expression for the energy in the coupling spring for a system such as that of problem 6.10.

Further Reading

Anderson, R.A., *Fundamentals of Vibrations* (Macmillan, New York, 1967)
Church, A.H., *Mechanical Vibrations*, 2nd edition (Wiley, New York, 1963)
Evan-Iwanowski, R.M., *Resonance Oscillations in Mechanical Systems* (Elsevier, Amsterdam, 1976)
French, A.P., *Vibrations and Waves* (Nelson, London, 1971)
Haberman, C.M., *Vibration Analysis* (Merrill, Columbus, Ohio, 1968)
Main, I.G., *Vibrations and Waves in Physics* (Cambridge University Press, Cambridge, 1978)
Morse, P.M., *Vibration and Sound*, 2nd edition (McGraw-Hill, New York, 1948)
Seto, W.W., *Theory and Problems of Mechanical Vibrations*, (Schaum, New York, 1964)
Steidel, R.F. Jr., *An Introduction to Mechanical Vibrations*, 2nd edition (Wiley, New York, 1979)
Timoshenko, S., Young, D.H. and Weaver, W. Jr., *Vibration Problems in Engineering*, 4th edition (Wiley, New York, 1974)
Vernon, J.B., *Linear Vibration Theory*, (Wiley, New York, 1967)

7 Non-discrete-element Mechanical Vibratory Systems

7.1 Introduction and Objectives

In this chapter the aim is to describe broadly a range of mechanical vibratory systems which do not consist of discrete distinguishable mass/spring/damper elements but which are linear systems.

After reading this chapter the student should be able to

(a) describe the oscillatory behaviour of a mass/spring system in which the spring possesses distributed mass;
(b) outline the operation of the compound pendulum in which system also the mass is distributed;
(c) describe the vibrations of a stretched string with distributed mass and flexibility;
(d) describe the transverse vibrations of a solid rod;
(e) outline the longitudinal vibrations of the solid rod;
(f) discuss the vibratory behaviour of a stretched circular membrane;
(g) describe the possible vibrations in a column of air;
(h) construct electrical analogues for these distributed mechanical vibratory systems.

7.2 Simple Harmonic Oscillator with Massy Spring

In the discussions in the earlier chapters of the mechanical harmonic oscillator, each component in the system was considered to be discrete, that is, the body of mass M was not deformed and had no spring constant and the spring of constant C had no mass. In practice, however, the mass of the spring M_s may not be negligible. Motion of the spring itself then involves kinetic as well as potential energy but not all parts of the spring oscillate with the same amplitude. In the spring represented in figure 7.1 the right-hand end moves exactly as the body while the left-hand end remains fixed.

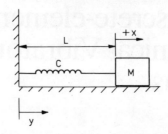

Figure 7.1 A simple harmonic oscillator in which the uniform spring of constant C has a length L and a mass M_s.

Assume that all parts of the spring move in phase. Then the velocity of the elemental portion of the spring at y, of length dy, and of mass dM_s, where

$$dM_s = \left(\frac{M_s}{L}\right) dy \qquad (7.1)$$

will be

$$\dot{x}_y = \left(\frac{y}{L}\right) \dot{x} \ . \qquad (7.2)$$

Here $\dot{x}$ is the velocity of the body. Thus the elemental kinetic energy will be

$$dE_{ks} = \tfrac{1}{2} (dM_s) \dot{x}_y^2$$

$$= \frac{M_s \dot{x}^2}{2L^3} y^2 \, dy \ . \qquad (7.3)$$

The total kinetic energy in the spring itself may be found thus

$$E_{ks} = \frac{M_s \dot{x}^2}{2L^3} \int_0^L y^2 \, dy$$

$$= \tfrac{1}{3} (\tfrac{1}{2} M_s \dot{x}^2) \ . \qquad (7.4)$$

So the total kinetic energy in the system, mass and spring, is given by

$$E_k = E_{km} + E_{ks}$$

$$= \tfrac{1}{2} M \dot{x}^2 + \tfrac{1}{2}\left(\tfrac{1}{3} M_s\right)\dot{x}^2$$

$$= \tfrac{1}{2}(M + \tfrac{1}{3} M_s)\dot{x}^2 \ . \qquad (7.5)$$

This actual system is therefore equivalent to a discrete component harmonic oscillator in which the spring has constant C and the body has an equivalent mass M_e where

$$M_e = M + \tfrac{1}{3}M_s \ . \tag{7.6}$$

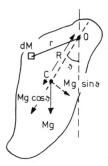

Figure 7.2 A compound pendulum which pivots about an axis through O and whose centre of gravity C is a distance R from O. The mass of the pendulum is M.

7.3 Compound Pendulum

A compound pendulum is a rigid body of arbitrary shape which can pivot about an axis through a point such as O in figure 7.2 other than its centre of gravity C. Let M be the total mass of the body and let R be the distance from O to C. At rest C lies vertically below O. If the body be rotationally deflected by an angle ϑ as shown there will be a restoring torque Tr such that

$$\text{Tr} = RMg \sin \vartheta \ . \tag{7.7}$$

For small values of ϑ

$$\text{Tr} = MgR\vartheta \ . \tag{7.8}$$

This torque will result in an angular acceleration $\ddot{\vartheta}$, related to the restoring torque thus from equation 1.17

$$MgR\vartheta = -I\ddot{\vartheta} \tag{7.9}$$

Here I is the mass moment of inertia of the body about the pivotal axis through O and is defined by

$$I = \int r^2 \, dM \tag{7.10}$$

where dM is an element of the mass at a distance r from the axis through O and the integration if over the whole body. Simplifying equation 7.9, by dividing each term by I, yields

$$\ddot{\vartheta} + \frac{MgR}{I}\vartheta = 0 \tag{7.11}$$

which is identical in form with equation 2.3, the equation of motion of the simple harmonic oscillator. The period T of the compound pendulum oscillation is

$$T = 2\pi \sqrt{\left(\frac{I}{MgR}\right)}. \tag{7.12}$$

Example 7.1

A uniform bar of length L m and total mass M kg hangs from a pivot at one end. A concentrated mass M kg may be attached to the bar anywhere along its length. By what percentage will the period of this compound pendulum decrease when the latter mass is moved from the lower extremity of the bar to the mid-point?

Solution

Let T_1 and T_2 be the first and second periods. I_b, the moment of inertia of the bar alone $= ML^2/3$. I_{m1}, the moment of inertia of the movable mass in the first case, $= ML^2$.

$$I_1 = I_b + I_{m1} = \frac{4}{3}ML^2.$$

If I_{m2} be the moment of inertia of the mass in the second instance, $I_{m2} = \frac{1}{4}ML^2$, and so

$$I_2 = I_b + I_{m2} = \frac{7}{12}ML^2.$$

In case 1, R_1 the distance from pivot to centre of gravity is $\frac{3}{4}L$. In case 2, $R_2 = \frac{1}{2}L$. Now from equation 7.12

$$\frac{T_2}{T_1} = \sqrt{\left(\frac{I_2 R_1}{R_2 I_1}\right)} = \sqrt{\left(\frac{7 \times 3 \times 2 \times 3}{12 \times 4 \times 1 \times 4}\right)} = 0.80 \text{ or 20 per cent down.}$$

NON-DISCRETE-ELEMENT MECHANICAL VIBRATORY SYSTEMS

Figure 7.3 A string of length L stretched between two fixed supports. The x-axis is taken along the string with the origin at the left-hand end.

7.4 Vibrating String

Consider the string of length L m stretched under a tensile force F N between two rigid supports. Take the x-axis along the string with $x = 0$ at the left-hand end. The set-up is shown in figure 7.3. Let the mass per unit length of the string be uniform κ kg/m such that

$$\kappa = \frac{M}{L} \tag{7.13}$$

where M is the mass of the string. If any part of the string is displaced from the horizontal, restoring forces come into play.

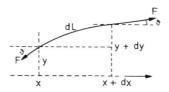

Figure 7.4 An elemental length of the string dL which has been deflected laterally in the y-direction. One end of the element has been deflected by y and the other end of the element by $y + dy$. The string makes an angle ϑ with the horizontal.

Let there be a vertical displacement of y at x and $y + dy$ at $x + dx$ as shown in figure 7.4. If these displacements are small, F may be taken as constant along the string. For the elemental length of string dL, the net vertical or y-component of F is the restoring force thus

$$dF_y = F \sin \vartheta(x + dx) - F \sin \vartheta(x) \tag{7.14}$$

where the two values of ϑ are not equal. $F \sin \vartheta(x + dx)$ may be expanded in a Taylor series

$$F \sin \vartheta(x + dx) = F \sin\vartheta(x) + \frac{\partial[F \sin \vartheta(x)]}{\partial x} dx + \ldots \quad (7.15)$$

Higher order terms would be very small and negligible. Now equation 7.14 becomes

$$dF_y = \frac{\partial}{\partial x}[F \sin \vartheta(x)] \, dx \, . \quad (7.16)$$

Since y is small, each value of ϑ must be small and so $\sin \vartheta = \tan \vartheta$. But $\tan \vartheta = \partial y/\partial x$. Therefore

$$dF_y = \frac{\partial}{\partial x}\left(F\frac{\partial y}{\partial x}\right) dx$$

$$= F\frac{\partial^2 y}{\partial x^2} dx \, . \quad (7.17)$$

The mass of this element of string is $\kappa \, dL = \kappa \, dx$ and so Newton's law allows the following equation to be written

$$\kappa \, dx \, \ddot{y} = F\frac{\partial^2 y}{\partial x^2} dx$$

or

$$\ddot{y} = \left(\frac{F}{\kappa}\right) \frac{\partial^2 y}{\partial x^2} \, . \quad (7.18)$$

This equation, describing transverse motion of the string, is a wave equation which relates transverse displacement y along the length or x-axis with time. If

$$\sqrt{\left(\frac{F}{\kappa}\right)} = c \quad (7.19)$$

so that

$$\ddot{y} = c^2 \frac{\partial^2 y}{\partial x^2} \, , \quad (7.20)$$

c m/s is the speed of propagation of the transverse disturbance along the string.

The general solution of equation 7.20 takes the form

$$y = f_1(ct - x) + f_2(ct + x) \quad (7.21)$$

where f_1 and f_2 are arbitrary functions. Figure 7.5 shows that the $f_1(ct - x)$ portion of this solution is a wave travelling from left to right, the forward

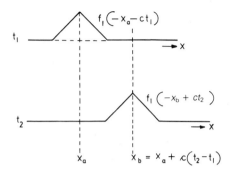

Figure 7.5 The propagation of a disturbance along the string in the x-direction. At t_1 the disturbance is at x_a and later, at t_2, the disturbance is at x_b. Thus the disturbance is a forward-travelling wave travelling at propagation speed c.

direction. It is a forward-travelling wave. The $f_2(ct + x)$ function represents a backwards or reverse travelling wave.

A harmonic solution to the wave equation is, in view of the Fourier theory, of considerable generality. Such a solution is

$$y = A_1 \sin(\omega t - \beta x) + A_2 \sin(\omega t + \beta x)$$
$$+ B_1 \cos(\omega t - \beta x) + B_2 \cos(\omega t + \beta x) \quad (7.22)$$

where A_1, A_2, B_1 and B_2 are arbitrary constants determined by the initial and boundary conditions and β is called the propagation or wavelength constant thus

$$\beta = \frac{\omega}{c}$$
$$= \frac{2\pi f}{c} . \quad (7.23)$$

If the terminal or boundary conditions of figure 7.3 apply, that is, $y = 0$ at $x = 0$ and $y = 0$ at $x = L$ then

$$0 = (A_1 + A_2) \sin \omega t + (B_1 + B_2) \cos \omega t \quad (7.24)$$

for all t. Therefore

$$A_1 + A_2 = 0$$

or

$$A_2 = -A_1 , \quad (7.25)$$

and

$$B_2 = -B_1. \tag{7.26}$$

Equation 7.22 then simplifies to

$$y = A_1 [\sin(\omega t - \beta x) - \sin(\omega t + \beta x)]$$
$$+ B_1 [\cos(\omega t - \beta x) - \cos(\omega t + \beta x)] . \tag{7.27}$$

Next, applying the boundary condition $y = 0$ at $x = L$, at all values of t including $t = 0$, yields

$$0 = A_1 [\sin(-\beta L) - \sin(\beta L)] + B_1 [\cos(-\beta L) - \cos(\beta L)]$$
$$= -2A_1 \sin(\beta L) . \tag{7.28}$$

Therefore

$$\sin(\beta L) = 0 \tag{7.29}$$

or

$$\beta L = n\pi \text{ for } n = 1, 2, 3 \ldots$$

and

$$\beta = \frac{n\pi}{L} . \tag{7.30}$$

Then

$$\omega = \frac{n\pi c}{L} \tag{7.31}$$

or

$$f = \frac{nc}{2L} \text{ for } n = 1, 2, 3 \ldots \tag{7.32}$$

Thus the string can sustain only a discrete set of frequencies in free vibration, the set of natural frequencies of the string.

The general solution of equation 7.22 is reduced to the following for $n = 1$, the first harmonic frequency or fundamental mode of the string

$$y_1 = 2(-A_1 \cos \omega_1 t + B_1 \sin \omega_1 t) \sin \beta_1 x \tag{7.33}$$

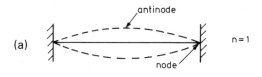

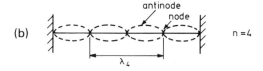

Figure 7.6 The standing wave patterns for free transverse vibrations (a) at the fundamental harmonic and (b) at the fourth harmonic frequency in the stretched string.

The nth harmonic or $(n-1)$th overtone has the solution

$$y_n = 2[-A_n \cos(n\omega_1 t) + B_n \sin(n\omega_1 t)] \sin(n\beta_1 x) \ . \qquad (7.34)$$

Each of these solutions is not a travelling wave but rather a standing wave or transverse vibration of each element of the string. The pattern of the vibrations for the fundamental and the fourth harmonics are shown in figure 7.6. At the nodes no lateral displacement occurs. At all other points sinusoidal vibrations occur with the maximum disturbance at the antinodes. The distance from one node to the next or from one antinode to the next is L/n. This is half the distance in which the vibration patterns of equations 7.33 and 7.34 repeat themselves. Thus y_n has the same value when x is increased by $2L/n$ in equation 7.34. This distance is called the wavelength λ where

$$\lambda_1 = 2L$$

$$= \frac{c}{f_1} \qquad (7.35)$$

and

$$\lambda_n = \frac{2L}{n}$$

$$= \frac{c}{f_n}$$

$$= \frac{c}{nf_1} \ . \qquad (7.36)$$

Example 7.2

A string of length 10 m of total mass 0.01 kg is drawn between two rigid supports by a tensile force of 5 N. Find the fundamental frequency for free vibrations of the string.

Solution

The mass per unit length $\kappa = M/L = 10^{-3}$ kg/m. Tensile force $F = 5$ N. So from equation 7.19 the propagation speed $c = 70.7$ m/s. Therefore the fundamental frequency, from equation 7.32, is

$$f_1 = \frac{c}{2L} = 3.58 \text{ Hz}.$$

7.5 Solid Rod in Transverse or Flexural Vibration

Consider the solid rod shown in figure 7.7 which is to experience bending distortion. The length of the rod is L m and the uniform cross-section has an area A m^2 and an axis of symmetry YY. Take the x-axis along the plane of symmetry of the rod with $x = 0$ at the left-hand end.

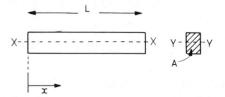

Figure 7.7 A rod of length L and cross-sectional area A. The area has a horizontal axis of symmetry YY. The plane through YY in the x-direction, XX, is the neutral axis of the rod. The x-axis is taken along this plane with the origin at the left-hand end of the rod.

When bent as shown in figure 7.8 the upper half of the cross-section of the bar is elongated the lower half is shortened and the plane of symmetry through YY and the x-axis is unchanged in length. This plane, labelled XX in figure 7.7, is called the neutral axis of the rod. Assume that the bending results in the neutral axis of a short length of the rod dx forming an arc of a circle of radius R as shown in figure 7.8. Consider the longitudinal lamina of the rod a distance h from the neutral axis and of cross-sectional area dA. It will be longer than the neutral axis dx by Δx where

NON-DISCRETE-ELEMENT MECHANICAL VIBRATORY SYSTEMS

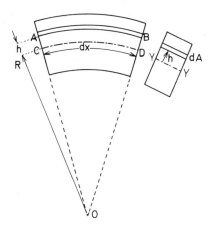

Figure 7.8 An elemental length of the rod dx is bent into the arc of a circle such that the neutral axis has a radius of curvature R. Elements of the bar above the neutral axis are lengthened while elements below are shortened. The neutral axis itself is unchanged in length.

$$\Delta x = \frac{\partial \xi}{\partial x} dx \ . \tag{7.37}$$

Here ξ is the lengthwise displacement of the particles of the rod. From figure 7.8, from the similar triangles ABO and CDO

$$\frac{dx + \Delta x}{R + h} = \frac{dx}{R} \tag{7.38}$$

and so

$$\frac{\Delta x}{dx} = \frac{h}{R}$$
$$= \frac{\partial \xi}{\partial x} \ . \tag{7.39}$$

These latter expressions are measures of the lengthwise strain in that laminar segment of the rod. Assuming that the material of the rod obeys Hooke's law of equation 1.22, and taking the Young's modulus for that material to be E, the elemental tensile force on that lamina will be

$$dF = E \, dA \, \frac{\partial \xi}{\partial x}$$
$$= \frac{Eh}{R} dA \ . \tag{7.40}$$

The total longitudinal force on the rod, however, must be zero since it is in equilibrium. Therefore the negative (compressive) forces below the neutral axis are counterbalanced by the positive (tensile) forces above but these forces do result in a bending moment M_B given by

$$M_B = \int h \, dF$$

$$= \frac{E}{R} \int h^2 \, dA . \qquad (7.41)$$

The expression $\int h^2 \, dA$ is the area moment of inertia of the cross-sectional area of the rod about the axis of symmetry YY. Denote it by I_A. Therefore

$$M_B = \frac{EI_A}{R} . \qquad (7.42)$$

Since R is not in general constant all along the length of the rod but rather is a function of x, the bending moment is also a function of x. If the vertical displacement y of the neutral axis of the bar upon bending is small, that is $\partial y/\partial x \ll 1$, then the radius of curvature is given approximately by

$$R = 1 \Big/ \left(\frac{\partial^2 y}{\partial x^2} \right) . \qquad (7.43)$$

The bending moment from equation 7.42, is given by

$$M_B = EI_A \frac{\partial^2 y}{\partial x^2} \qquad (7.44)$$

Besides the bending moments acting on the short segment of the rod as shown in figure 7.9, shear forces also act on it. Thus $F_y(x)$ acts at the

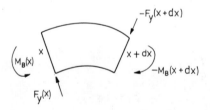

Figure 7.9 The shear forces F_y and bending moments M_B acting on the bent element of rod.

left-hand end of the segment x and $F_y(x + dx)$ acts at the right-hand end $x + dx$. The total turning moment or torque about the left-hand end of the segment is therefore

$$M_B(x) - M_B(x + dx) - F_y(x + dx) \, dx = 0 \qquad (7.45)$$

NON-DISCRETE-ELEMENT MECHANICAL VIBRATORY SYSTEMS

because the segment is under static equilibrium. For dx very small, both $M_B(x + dx)$ and $F_y(x + dx)$ may be expanded in Taylor series and only the first two terms retained as significant thus

$$M_B(x + dx) = M_B(x) + \frac{\partial M_B}{\partial x} dx \ldots \quad (7.46)$$

and

$$F_y(x + dx) = F_y(x) + \frac{\partial F_y}{\partial x} dx \ldots \quad (7.47)$$

Then equation 7.45 becomes

$$-\frac{\partial M_B}{\partial x} dx - F_y(x) \, dx - \left(\frac{\partial F_y}{\partial x} dx\right) dx = 0. \quad (7.48)$$

The second order term in $(dx)^2$ may be neglected and so

$$F_y = -\frac{\partial M_B}{\partial x}$$

$$= -EI_A \frac{\partial^3 y}{\partial x^3}. \quad (7.49)$$

This relationship holds for static equilibrium. In the dynamic state the total turning moment or torque produces angular acceleration of the segment about the left-hand end. For small vertical displacements this torque is always close to zero. Then the force F_y in equation 7.49 is very close to the accelerating force in the y- or vertical direction.

The net vertical force on the segment of length dx is given by

$$dF_y = F_y(x) - F_y(x + dx)$$

$$= -\frac{\partial F_y}{\partial x_4} dx$$

$$= EI_A \frac{\partial^4 y}{\partial x^4} dx. \quad (7.50)$$

This force causes vertical acceleration and so from Newton's second law

$$EI_A \frac{\partial^4 y}{\partial x^4} dx = (\rho A dx)\ddot{y} \quad (7.51)$$

where ρ kg/m³ is the density of the rod material. Therefore

$$\ddot{y} = \left(\frac{EI_A}{\rho A}\right) \frac{\partial^4 y}{\partial x^4} \quad (7.52)$$

This is the wave equation for the transverse vibration of the rod. In form, it is different from equations 7.18 or 7.20 for the string. Functions such as $f(ct \pm x)$ are not solutions and the transverse waves do not travel along the rod at constant speed. In fact, as will be seen below, each frequency component in the wave travels at its own speed. This wave is therefore said to be dispersive.

It can be shown that the following is a solution to the dispersive wave equation 7.52,

$$y = \cos(\omega t + \gamma)\left(A \cosh\frac{\omega x}{c} + B \sinh\frac{\omega x}{c}\right.$$
$$\left. + C \cos\frac{\omega x}{c} + D \sin\frac{\omega x}{c}\right). \quad (7.53)$$

Here A, B, C, D and γ are constants set by the initial conditions and boundary conditions of the system. Also

$$c^2 = \omega\sqrt{\left(\frac{EI_A}{\rho A}\right)} \quad (7.54)$$

where c is the phase velocity of the harmonic of frequency f ($= \omega/2\pi$).

To find the values of the amplitude constants and of the phase angle γ, knowledge of a number of boundary conditions must be available. The conditions at the boundaries, in this case the ends of the rod at $x = 0$ and at $x = L$, can enforce strict relationships between the values of the amplitude constants. If an end of the bar is rigidly clamped then at that end $y = 0$ and $dy/dx = 0$, the transverse deflection and the slope are held to zero there. If an end is free, no bending moment or shear force can exist there and so from equations 7.44 and 7.49 at that end $d^2y/dx^2 = 0$ and $d^3y/dx^3 = 0$.

Consider the case of a bar free at both ends similar to a tuning fork. At $x = 0$, $d^2y/dx^2 = 0$ and $d^3y/dx^3 = 0$. By inserting these conditions into equation 7.53, one finds that $A = C$ and $B = D$. So there can be only two independent amplitude constants. Proceeding further and applying the same criteria at $x = L$, it is found that A and B are non-zero only on condition that

$$\cosh\left(\frac{\omega L}{c}\right)\cos\left(\frac{\omega L}{c}\right) = 1 \quad (7.55)$$

or

$$\tan\left(\frac{\omega L}{2c}\right) = \pm\tanh\left(\frac{\omega L}{2c}\right). \quad (7.56)$$

This equation holds only at certain discrete frequencies. Therefore only at these frequencies can transverse or bending waves be sustained in the rod. These frequencies follow from the solutions to equation 7.56, thus

$$\frac{\omega L}{2c} = \frac{\pi}{4}(3.0112, 5, 7, 9 \ldots) \quad (7.57)$$

and are

NON-DISCRETE-ELEMENT MECHANICAL VIBRATORY SYSTEMS

$$f = \frac{\pi}{8L^2}\sqrt{\left(\frac{EI_A}{\rho A}\right)}(9.067, 25, 47, 81 \ldots) \ . \tag{7.58}$$

Note that these overtone frequencies are not integral factors of the lowest or fundamental.

The final three independent constants in equation 7.53, describing the transverse wave in the rod, A, B and γ, can be determined from three independent initial conditions.

Example 7.3

A steel rod 1 m long × 1 cm × 1 cm is mounted as a cantilever. Given that Young's modulus for steel is 0.2×10^{12} N/m^2, find the speed of propagation of transverse waves along the rod. The density of steel is 7.8×10^3 kg/m^3.

Solution

The area moment of inertia about neutral axis is

$$I_A = (10^{-2})^4/24 = 4.17 \times 10^{-10} \text{ m}^4 \ .$$

Here $c^2 = \omega\sqrt{(EI_A/\rho A)}$

$$= 10.34 \ \omega$$

$$= 64.97f \ .$$

Then $c = 8.06 \ \sqrt{(f)}$ m/s .

7.6 Longitudinal Vibrations in a Solid Rod

The rod of figure 7.7, may also be subjected to lengthwise or longitudinal forces which cause longitudinal displacement ξ of the particles of the bar. For a long bar of small cross-sectional area A, all particles in each cross-section experience the same displacement.

Consider the elemental portion of the rod between x and $x + dx$ in figure 7.10. The longitudinal displacement changes from ξ at x to $\xi + d\xi$ at $x + dx$ where $d\xi$ may be found from the second term of the Taylor series expansion of ξ with x

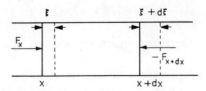

Figure 7.10 An elemental length of the rod between x and $x + dx$ undergoing longitudinal strain. The cross-section at x is displaced by ξ and the cross-section at $x + dx$ is displaced by $\xi + d\xi$.

$$d\xi = \left(\frac{\partial \xi}{\partial x}\right) dx \; . \tag{7.59}$$

The strain of the segment of the rod is S_x which is given by

$$\begin{aligned} S_x &= \frac{(\partial \xi/\partial x) \, dx}{dx} \\ &= \frac{\partial \xi}{\partial x} \; . \end{aligned} \tag{7.60}$$

Assuming elastic behaviour, the stress in the bar T_x may be calculated, if Young's modulus of the material of the bar E is known, thus

$$\begin{aligned} T_x &= \frac{F_x}{A} \\ &= ES_x \\ &= E\frac{\partial \xi}{\partial x} \end{aligned} \tag{7.61}$$

where the convention of compressive force being taken as negative and tensile force as positive is adopted. Therefore

$$F_x = AE\frac{\partial \xi}{\partial x} \; . \tag{7.62}$$

The element of the bar in figure 7.10 experiences a net force which accelerates it. This unbalanced force is

$$\begin{aligned} F_x - F_{x+dx} &= F_x - F_x - \frac{\partial F_x}{\partial x}dx \\ &= -\frac{\partial F_x}{\partial x} dx \\ &= -AE\frac{\partial^2 \xi}{\partial x^2}dx \\ &= -\rho A \, dx \, \ddot{\xi} \; . \end{aligned} \tag{7.63}$$

NON-DISCRETE-ELEMENT MECHANICAL VIBRATORY SYSTEMS

This equation may be simplified into the wave equation thus

$$\ddot{\xi} = \frac{E}{\rho}\frac{\partial^2 \xi}{\partial x^2}$$

$$= c^2 \frac{\partial^2 \xi}{\partial x^2}. \tag{7.64}$$

Equation 7.64 is identical in form with the wave equation for the stretched string of equation 7.20. In this case, however, the propagation speed c is given by

$$c = \sqrt{\left(\frac{E}{\rho}\right)} \tag{7.65}$$

and is constant at all frequencies. The longitudinal wave is thus a non-dispersive wave in contrast with the transverse wave in the same rod.

The solution to this wave equation takes the forms given in equations 7.21 and 7.22. The boundary conditions at $x = 0$ and at $x = L$ determine the relationships between the amplitude constants. Take, for example, the case previously considered for transverse vibrations in which the two ends of the rod are free. This means that at $x = 0$, $F_x = 0$ or, from equation 7.62 $\partial \xi / \partial x = 0$ and, at $x = L$, $\partial \xi / \partial x = 0$. Substituting in equation 7.22, at $t = 0$

$$(-\beta A_1 + \beta A_2) \cos \omega t + (\beta B_1 - \beta B_2) \sin \omega t = 0$$

and so

$$A_2 = A_1 = A$$

and

$$B_2 = B_1 = B.$$

Carrying out the same substitution at $x = L$ produces

$$-A \cos(\omega t - \beta L) + A \cos(\omega t + \beta L) + B \sin(\omega t - \beta L) - B \sin(\omega t + \beta L) = 0.$$

If A and B are non-zero then

$$\sin \beta L = 0 \tag{7.66}$$

and

$$\beta L = n\pi \text{ for } n = 1, 2, 3 \ldots \tag{7.67}$$

or

$$f = \frac{n}{2L}\sqrt{\left(\frac{E}{\rho}\right)} \quad \text{for } n = 1, 2, 3 \ldots \tag{7.68}$$

This set of discrete frequencies constitutes the only frequencies at which longitudinal waves can travel along the rod when it is freely vibrating. Note that they are harmonically related.

The lengthwise displacement ξ as a function of location x and time t from equation 7.22 becomes

$$\xi = A[\sin(\omega t - \beta x) + \sin(\omega t + \beta x)]$$

$$+ B[\cos(\omega t - \beta x) + \cos(\omega t + \beta x)]$$

$$= 2(A \sin \omega t + B \cos \omega t) \cos \beta x$$

$$= 2\sqrt{(A^2 + B^2)} \cos\left[\omega t - \tan^{-1}\left(\frac{A}{B}\right)\right] \cos \beta x . \tag{7.69}$$

Note that there are in this instance only two independent parameters, $\sqrt{(A^2 + B^2)}$ and the phase angle $\tan^{-1}(A/B)$, to be determined from the initial conditions. Equation 7.69 is the equation for a standing wave along the rod. At any location x on the rod the amplitude of the longitudinal vibration is a constant $2\sqrt{(A^2 + B^2)} \cos \beta x$. Also only discrete overtones of the fundamental frequency $(1/2L)\sqrt{(E/\rho)}$ Hz can be sustained in free vibration in the bar. Figure 7.11 illustrates the variation in amplitude of this displacement along the length of the bar for three harmonics, $n = 1, 2$ and 4 for a hypothetical bar of $\sqrt{(E/\rho)} = 1$. In practice, each harmonic would have its own initial conditions not necessarily the same as each other.

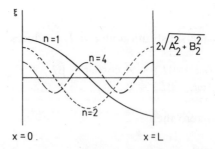

Figure 7.11 The standing wave patterns for free longitudinal waves in the rod at the fundamental ($n = 1$), second harmonic ($n = 2$) and fourth harmonic ($n = 4$) frequencies.

NON-DISCRETE-ELEMENT MECHANICAL VIBRATORY SYSTEMS

Other types of loading on the ends of the rod would give rise to different relations between the amplitude constants and hence to other patterns of standing waves and other sustainable frequencies in free oscillation.

Example 7.4

What is the speed of propagation of longitudinal waves in the steel rod of example 7.3? What fundamental frequency of this form of free vibration is sustained in the rod?

Solution

From equation 7.65

$$c = \sqrt{\left(\frac{E}{\rho}\right)} = 5060 \text{ m/s}$$

From equation 7.68

$$f_{(n=1)} = \frac{c}{2L} = 2.53 \text{ kHz} .$$

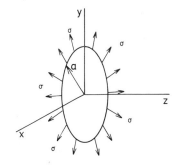

Figure 7.12 A thin circular membrane of radius a mounted in the xy plane with a tensile force σ per unit length of periphery. The z-axis is perpendicular to the membrane with the origin at the centre.

7.7 Vibrating Circular Membrane

A thin stretched membrane is another system with distributed mass and spring constant that can engage in vibrations. Consider such a uniform circular membrane of radius a m, mass per unit area μ kg/m² and tensile force per unit length of periphery σ N/m. Take the x and y-axes in the plane of the membrane and the z-axis perpendicular to the centre of it as shown in figure 7.12. The membrane bulges along the z-axis. When it does so bulge, restoring forces due to the tensile force on the membrane come into play.

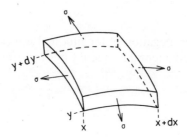

Figure 7.13 An elemental portion of the thin membrane between x and $x + dx$ and y and $y + dy$ which is deflected in the z-direction.

For the elemental portion of the membrane shown in figure 7.13, the net restoring force along the x-axis is

$$\sigma dy \left[\left(\frac{\partial z}{\partial x} \right)_{x+dx} - \left(\frac{\partial z}{\partial x} \right)_{x} \right] = \sigma \frac{\partial^2 z}{\partial x^2} dy dx \tag{7.70}$$

where, as in equation 7.9, a Taylor series expansion of $(\partial z/\partial x)_{x+dx}$ is used. Likewise the net force along the y axis is $\sigma(\partial^2 z/\partial y^2) dy dx$. These two forces accelerate the mass of the membrane element thus

$$\sigma \left(\frac{\partial^2 z}{\partial x^2} + \frac{\partial^2 z}{\partial y^2} \right) dx\, dy = \mu\, dx\, dy\, \ddot{z} \tag{7.71}$$

so that the following equation may be written

$$\ddot{z} = \frac{\sigma}{\mu} \left(\frac{\partial^2 z}{\partial x^2} + \frac{\partial^2 z}{\partial y^2} \right)$$
$$= c^2 \nabla^2 z\ . \tag{7.72}$$

Here ∇^2 is the two dimensional Laplacian operator defined by this equation. This is a two dimensional wave equation where the wave is non-dispersive since the speed of propagation is independent of frequency thus

$$c = \sqrt{\left(\frac{\sigma}{\mu} \right)}\ . \tag{7.73}$$

The wave equation will have forward-moving solutions of the general form

NON-DISCRETE-ELEMENT MECHANICAL VIBRATORY SYSTEMS

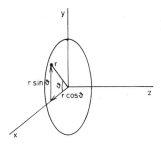

Figure 7.14 Cylindrical co-ordinates r, ϑ, z with the origin at the centre of the circular membrane and the membrane mounted in the $r\vartheta$ plane.

$$z = f(ct - x \cos \vartheta - y \sin \vartheta) \tag{7.74}$$

indicating wave propagation along a direction on the membrane making an angle ϑ with the x-axis.

Equation 7.72 may be readily converted to cylindrical co-ordinates (r, ϑ, z) which are more convenient for the circular membrane. In these co-ordinates, illustrated in figure 7.14, equation 7.72 converts to

$$\ddot{z} = c^2 \left(\frac{\partial^2 z}{\partial r^2} + \frac{1}{r} \frac{\partial z}{\partial r} + \frac{1}{r^2} \frac{\partial^2 z}{\partial \vartheta^2} \right). \tag{7.75}$$

If circular symmetry applies, z is independent of ϑ and equation 7.75 reduces to

$$\ddot{z} = c^2 \left(\frac{\partial^2 z}{\partial r^2} + \frac{1}{r} \frac{\partial z}{\partial r} \right). \tag{7.76}$$

A solution can be found for this wave equation by the method of separation of variables and this solution is the product of a function of r, $A(r)$, and a harmonic function of time t, $\cos(\omega t + \gamma)$ thus

$$z = A(r) \cos(\omega t + \gamma). \tag{7.77}$$

Substituting this solution into equation 7.76 yields

$$\frac{d^2 A(r)}{dr^2} + \frac{1}{r} \frac{dA(r)}{dr} + \beta^2 A(r) = 0 \tag{7.78}$$

where

$$\beta = \frac{\omega}{c}$$

is the radial wavelength constant or propagation constant. Equation 7.78 is a form of Bessel's differential equation

$$\frac{d^2 z}{dr^2} + \frac{1}{r}\frac{dz}{dr} + \left(1 - \frac{m^2}{r^2}\right)z = 0 \qquad (7.79)$$

the solutions of which are published and tabulated as the Bessel functions. One solution to equation 7.78 is

$$A(r) = AJ_0(\beta r) \qquad (7.80)$$

where A is an arbitrary constant and $J_0(\beta r)$ is the Bessel function of βr of the first kind and zero order. This latter function is illustrated in figure 7.15. $N_0(\beta r)$, the Bessel function of the second kind and zero order, is also a solution to equation 7.78. Since this function becomes infinite at $r = 0$, however, it can not satisfy the observed conditions on a practical membrane and so it can be neglected.

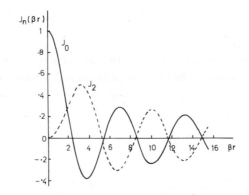

Figure 7.15 Bessel functions of the first kind of zero order J_0 and of second order J_2 of the argument βr.

The full solution to equation 7.76 then is

$$z = AJ_0(\beta r) \cos(\omega t + \gamma) \ . \qquad (7.81)$$

The boundary conditions determine the allowable values of f or ω or β. Thus, if the circular membrane is rigidly tethered around the circumference, at $r = a$, $z = 0$, then

$$J_0(\beta a) = 0$$

and $\beta a = 2.41, 5.52, 8.65, 11.79 \ldots$ \hfill (7.82) from figure 7.15. This latter equation specifies the allowable frequencies for free vibration of the membrane

$$f_1 = \frac{2.41c}{2\pi a}, \text{ the fundamental frequency}$$

$$f_2 = \frac{5.52c}{2\pi a}, \text{ the first overtone}$$

$$f_3 = \frac{8.65c}{2\pi a}, \text{ the second overtone, etc.} \qquad (7.83)$$

These are some of the allowed frequencies for free vibration of the membrane where circular symmetry holds. Some of these modes of vibration are shown in figure 7.16 (a). Note that they are not harmonically related to each other at all.

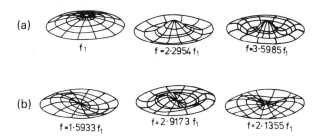

Figure 7.16 (a) Some of the circularly symmetrical modes of free vibration of the thin membrane. (b) Some of the low frequency free vibration modes of the membrane when circular symmetry does not hold. (After Morse).

The complete solution to the wave equation of equation 7.76 would take the form

$$z = \Sigma A_n \cos(\omega_n t - \gamma_n) J_0(\beta_n r) . \qquad (7.84)$$

The amplitude constants A_n and the phase angles γ_n would be determined by the initial conditions of the membrane.

If circular symmetry did not apply in the vibration and if equation 7.75 were the wave equation to be solved, still more complex modes of motion of the membrane would then be allowable. Figure 7.16 (b) illustrates some of these modes in the lower frequency range.

Example 7.5

A steel diaphragm 10^{-5} m thick and of 10^{-2} m radius is stretched to a tensile force per unit length of periphery 10^5 N/m. The density of steel is 7.8×10^3 kg/m³. Find the speed of radial propagation of transverse vibrations in the diaphragm. Calculate the fundamental frequency of sustained free circularly symmetrical vibrations.

Solution

Tensile force per length of circumference $\sigma = 10^5$ N/m. Mass per unit area $\mu = \rho \times$ thickness $= 7.8 \times 10^{-2}$ kg/m². From equation 7.73, the radial propagation speed, $c = \sqrt{(\sigma/\mu)} = 1130$ m/s. The fundamental frequency, f_1, if found from equation 7.83

$$f_1 = \frac{2.41c}{2\pi a}$$

$$= 43.3 \text{ kHz}.$$

7.8 Vibrations in Air Columns

Longitudinal vibrations may be readily set up in an air column, a rigid tube of uniform cross-sectional area A m² containing air, as in figure 7.17. The wave equation for this case, for an air medium, has a derivation and a form identical with equation 7.64.

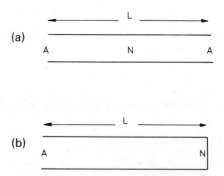

Figure 7.17 (a) A column of air of length L open at both ends. The standing wave pattern will have antinodes at the two open ends and a node at the mid-point. (b) If one end of the column of air is closed and the other open, a node will exist in the standing wave pattern at the closed end and an antinode will exist at the open end.

Some points in regard to the speed of propagation of the longitudinal vibrations in air should be noted. Equation 7.65 describing the speed in a solid

$$c = \sqrt{\left(\frac{E}{\rho}\right)}$$

must be modified for a fluid such as air because the stress or pressure in a fluid acts equally in all directions. Young's modulus E must be replaced by the bulk modulus B which is defined thus

$$B = -\frac{\partial P}{(\partial V/V)}$$

$$= -V\frac{\partial P}{\partial V} \quad . \tag{7.85}$$

the change in pressure divided by the resultant fractional change in volume. For most frequencies of interest, the vibratory changes in the air are adiabatic in that no heat flows during those changes. The pertinent bulk modulus is therefore the adiabatic bulk modulus B_{ad}. Thus the speed of propagation in air is given by

$$c = \sqrt{\left(\frac{B_{ad}}{\rho}\right)} . \tag{7.86}$$

In an ideal gas, to which air is an approximation, the pressure P and the volume V are related during an adiabatic process by

$$PV^\gamma = k \tag{7.87}$$

where k is a constant and γ is the ratio of the specific heat capacity of the gas at constant pressure to the specific heat capacity at constant volume. γ is a constant for each gas under given ambient conditions. For an adiabatic process, therefore, using equation 7.87

$$B_{ad} = -V\left(-\frac{\gamma P}{V}\right)$$

$$= \gamma P \quad . \tag{7.88}$$

For an ideal gas the density is given by

$$\rho = \frac{MP}{RT} \tag{7.89}$$

where M is the molecular mass, R is the universal gas constant and T is the absolute temperature in K.

Thus the speed of propagation c in such a gas is given by

$$c = \sqrt{\left(\frac{\gamma P}{\rho}\right)}$$

$$= \sqrt{\left(\frac{\gamma RT}{M}\right)}$$

$$= k_1 \sqrt{T} \, . \tag{7.90}$$

Here k_1 is a constant for each gas. For air at normal atmospheric pressure this propagation speed is a function of temperature $t\ °C$ thus

$$c = 331.6 + 0.6t \text{ m/s} \, . \tag{7.91}$$

The solutions to the wave equation in the air column take the form of equation 7.22. The boundary conditions, especially at $x = 0$ and at $x = L$ help to determine relationships between the amplitude terms A_1, A_2, B_1 and B_2 as well as the allowed frequencies for free vibrations in the air column.

If the tube is open at both ends as in figure 7.17(a) then an antinode is always positioned at each open end. The allowed frequencies are then

$$f_n = \frac{nc}{2L} \text{ for } n = 1, 2, 3 \ldots \tag{7.92}$$

If the tube is open at one end and closed at the other, as in figure 7.17(b), a node is always located at the closed end and an antinode at the open end. The allowed frequencies are then

$$f_m = \frac{(2m-1)c}{4L} \text{ for } m = 1, 2, 3 \ldots \tag{7.93}$$

In both cases the allowed frequencies are harmonically related sets. These latter two equations show that the frequencies sustainable in an air column, such as an organ pipe or tin whistle, are set by the length of the column and by whether the ends are open or closed. In the case of an open end, some air beyond the end is also involved in the oscillatory longitudinal motion. A correction must be applied to the column length in this case. This additional length ΔL is given as follows for a circular column of cross-sectional area A m^2:

$$\Delta L = \frac{8}{3\pi} \sqrt{\left(\frac{A}{\pi}\right)}$$

$$= 0.48 \sqrt{(A)} \, . \tag{7.94}$$

Consequently equation 7.92, for a column with two open ends, becomes

NON-DISCRETE-ELEMENT MECHANICAL VIBRATORY SYSTEMS

$$f_n = \frac{nc}{2(L + 2\Delta L)} \ . \tag{7.95}$$

Likewise, equation 7.93, for a column open at one end and closed at the other, becomes

$$f_m = \frac{(2m - 1)c}{4(L + \Delta L)} \ . \tag{7.96}$$

Example 7.6

Find the fundamental frequency at 20 °C of a pipe 0.3 m long open at both ends. The cross-sectional area of the circular pipe is 4 cm². What if the pipe is closed at one end?

Solution

End correction for the length, from equation 7.94

$$\Delta L = 0.48\sqrt{(A)}$$

$$= 0.0096 \text{ m} \ .$$

Corrected length $= L + 2\Delta L = 0.3192$ m. At 20 °C, $c = 331.6 + 12.0 = 343.6$ m/s. From equation 7.95, the fundamental frequency

$$f_1 = \frac{343.6}{2 \times 0.3192} = 538.2 \text{ Hz} \ .$$

In the second case the corrected length is $L + \Delta L = 0.3096$ m. From equation 7.96

$$f_1 = \frac{c}{4(L + L)} = 277.5 \text{ Hz} \ .$$

7.9 Electrical Analogue of Distributed Oscillators

A high-frequency transmission line is a direct analogue of any one of the non-dispersive continuum vibratory systems described in this chapter, such as the stretched string, the solid rod in longitudinal vibration and the air column. Such a line has a distributed series inductance L H per metre and distributed parallel capacitance C F per metre. An elemental segment of

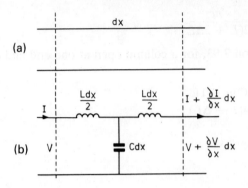

Figure 7.18 A high-frequency transmission line is a good analogue of a non-dispersive continuum mechanical vibratory system. An element of length of such a transmission line dx in (a) may be represented by the T-circuit in (b). C is the parallel capacitance per metre and L is the series inductance per metre.

such a line of length dx may be represented by the equivalent circuit of figure 7.18.

The voltage change across dx is given by

$$\frac{\partial V}{\partial x} dx = (-L\, dx)\, \dot{I}$$

or

$$\frac{\partial V}{\partial x} = -L\dot{I}. \tag{7.97}$$

The decrease in current across dx is

$$\frac{\partial I}{\partial x} dx = -C\, dx\, \dot{V}$$

or

$$\frac{\partial I}{\partial x} = -C\dot{V}. \tag{7.98}$$

Now differentiate equation 7.97 with respect to x and equation 7.98 with respect to t to find that

$$\frac{\partial^2 V}{\partial x^2} = -L\frac{\partial \dot{I}}{\partial x} \tag{7.99}$$

NON-DISCRETE-ELEMENT MECHANICAL VIBRATORY SYSTEMS

and

$$\frac{\partial I'}{\partial x} = -CV''. \qquad (7.100)$$

The order of partial differentiation does not matter and so from equations 7.99 and 7.100

$$V'' = \frac{1}{LC}\frac{\partial^2 V}{\partial x^2}$$

$$= c^2 \frac{\partial^2 V}{\partial x^2} \qquad (7.101)$$

and

$$I'' = \frac{1}{LC}\frac{\partial^2 I}{\partial x^2}$$

$$= c^2 \frac{\partial^2 I}{\partial x^2}. \qquad (7.102)$$

These are both wave equations identical in form with equation 7.20 and both with speed of propagation

$$c = \frac{1}{\sqrt{(LC)}} \qquad (7.103)$$

The solutions to these equations can take the forms given in equations 7.21 and 7.22. The relationships between the amplitude parameters can be derived from the boundary conditions, typically the current and voltage values at $x = 0$, the input to the transmission line and, at $x = L$, the output or termination of the line.

The introduction of distributed series resistance, R Ω per metre, can readily encompass damping in the mechanical system.

The corresponding analogous quantities and parameters for the stretched string, the longitudinally vibrating solid rod and the electrical transmission line are listed in table 7.1.

7.10 Experiments

(a) Vertical Spring Oscillator with Massy Spring

Measure the mass of the spring and use an appended mass not much larger than this. Check equation 7.6 by measuring the period accurately and showing that $T = 2\pi\sqrt{(M_e/C)}$ where M_e is given by equation 7.6.

Table 7.1 Analogies between certain Vibrating Systems: Stretched String in Transverse Vibration, Solid Rod in Longitudinal Vibration, Electrical Transmission Line

String	Rod	Transmission Line
mass per metre ($\varkappa$)	density (ρ)	inductance per m (L)
tensile force (F)	Young's modulus (E)	$\dfrac{1}{\text{capacitance per m}}\left(\dfrac{1}{C}\right)$
Damping per metre		resistance per m (R)
particle displacement (ξ)	particle displacement (ξ)	charge (q)
particle velocity ($\dot{\xi}$)	particle velocity ($\dot{\xi}$)	current (i)

(b) Compound Pendulum

A convenient form of compound pendulum is a metal bar of rectangular cross-section about a metre long and with holes drilled at regular intervals along its length. These holes can serve as support points for a knife-edge pivot. The mass of the bar is first measured. When the bar is mounted as a pendulum with the pivot in one of the holes a distance R from the centre of the bar, the period of the oscillation is accurately measured and is equal to $T = 2\pi\sqrt{(I/MgR)}$. R may be varied by choosing a different pivot hole. A graph of T^2 versus $1/R$ produces a straight line of slope $4\pi^2 I/Mg$. If g is known then I may be calculated and conversely if I is known g can be determined.

(c) Sonometer Experiments

A sonometer consists of a wire stretched between two support points under a variable tensile force. This force is achieved by passing the wire over a pulley and hanging weights on the end of the wire. The distance between the two supports may be varied also. The total mass of the wire is measured and the total length in order to calculate the mass per unit length κ. A sound of known frequency is generated by supplying an appropriate voltage from a signal generator to a loudspeaker. For a given tensile force in the wire of the sonometer, the length is varied until the pitch of the sound, when the wire is plucked at the centre-point, corresponds to that of the sound from the loudspeaker. Then $f = \sqrt{(F/\kappa)}/2L$ where F is the tensile force and L the length between the supports. If F is increased and L determined for the same frequency at each F value, a graph of F versus L^2

is a straight line of slope $4f^2\kappa$. The same experiment can be repeated at a number of different frequencies and equation 7.32 verified.

(d) Transverse Beam Oscillations

A tuning fork is basically a beam tethered at the mid-point and with the two ends free. Measure the frequency of the fundamental for a number of tuning forks made of the same material. Measure the dimensions of the beam and check equation 7.58.

(e) Resonance Tube

The resonance tube is an open-ended glass tube which can be stuck into the surface of a water reservoir so that the length of the air column in it can be varied. The air column can be made to vibrate longitudinally by placing a vibrating tuning fork close to the open end. When the length of the air column (plus the end correction) corresponds to a quarter wavelength of the frequency of the tuning fork sound, the air column itself is noticeably a sound generator or resonates. Then the speed of sound in air, $c = 4fL$, may be calculated where f is the frequency and L the corrected air column length. If a number of different frequencies are used and the corresponding corrected lengths measured, a graph of f versus $1/L$, is a straight line of slope $c/4$. Therefore c may be found.

(f) Kundt's Tube

Kundt's tube is a glass tube filled with air with one end tightly closed and the other end loosely closed with a cork on the end of a brass rod. The mid-point of the rod is rigidly clamped. The inside of the tube is thoroughly dried out and a light dusting of lycopodium powder is placed inside. The brass rod is caused to resonate by stroking it with a resined cloth. At the fundamental resonance of the rod, the length is one half-wavelength. As the rod resonates, the position of the end of the rod in the tube is varied until resonance is obtained in the air column as well. Then the powder accumulates at the nodes of the standing wave in the air column. The distance from one node to the next can be measured and is one half-wavelength for that frequency in air. The speed of sound in air can be calculated $c = f\lambda$ and $f = (1/2L_r)\sqrt{(E/\rho)}$, where L_r is the length of the rod, E the Young's modulus of brass and ρ the density of brass.

Alternative forms of Kundt's tube have electrically driven loudspeakers as the sound source. In such an instrument the frequency may be measured on an oscilloscope or with an electronic frequency meter and for a given length of air column in the tube the frequency can be varied until resonance is obtained.

(g) Drum Membrane Modes

The modes in a vibrating membrane can be studied in a fashion very similar to that used in the Kundt's tube. The membrane is positioned horizontally and a light powder dusted on the top. The natural frequency transverse vibrations of the membrane may be generated by lightly tapping the underside of the membrane at different radial distances. Alternatively a loudspeaker driven by a variable frequency signal generator can be placed under the membrane and the resonances stimulated at the frequencies predicted by equation 7.83 and indicated in figure 7.16. The speed of propagation of the vibration in the membrane may be calculated from this equation.

The radial tension in the membrane may be varied or else other membranes, either of different mass per unit area or of different radii, may also be investigated.

(h) Transmission Lines

Short lengths of transmission lines may be set up in the laboratory. Wires of different materials and gauges positioned at different distances apart allow a degree of latitude in the values of resistance per unit length, inductance per unit length and capacitance per unit length. The input A.C. voltage is connected to one end and the output end may be terminated in any way desired with open circuit, short circuit or any impedance. A voltage detector placed across between the two wires allows the voltage difference between the wires to be measured at any location. The standing wave pattern along the wires may thus be drawn and the wavelength determined (= four times the distance from a node to an antinode). The standing wave pattern for different frequencies and for different terminations provides useful insights into the distributed parameter vibrating system.

Problems

7.1 A stretched string of length 1 m produces a fundamental frequency of 500 Hz. The total mass of the string is 0.01 kg. Find the tensile force on the string.

7.2 A forward-travelling wave on a string is described by $y = 0.001 \cos(5t + 4z)$. What are the frequency, propagation speed and wavelength? What are the transverse velocity and acceleration amplitudes?

7.3 A mass of 5 kg hangs on a spring of mass 2 kg and spring constant 200 N/m. There is damping of coefficient 5 kg/s. Calculate the following features of this system: period of damped free vibration, Q-value, mechanical impedance at 0.5 Hz (magnitude and phase angle), r.m.s. velocity of the mass when driven by a sinusoidal force of amplitude 10 N and frequency 0.5 Hz and average power dissipation in the system in this latter case.

7.4 A uniform bar of length 2 m and mass 5 kg hangs from a pivot at one end. During side-to-side oscillation, the amplitudes of successive swings decline by a factor of 0.8. Find the period of the damped and the undamped oscillations. What is the effective damping ratio?

7.5 The distance between two nodes on a string of 0.001 kg/m is 0.32 m for a frequency of 35 Hz. Find the speed of propagation of the wave and the tensile force on the string.

7.6 The fundamental frequency of a stretched wire of length 80 cm is 100 Hz. The peak-to-peak excursion of the wire at the mid-point is 4 cm. The total mass of the wire is 0.04 kg. Calculate the speed of propagation of the wave, the tensile force in the wire and the transverse velocity of the wire 20 cm from one end.

7.7 Find the amplitude and phase angle of the travelling wave $y = 0.02 \cos(5t - 2z) + 0.035 \sin(5t - 2z)$. What are the frequency and the speed of propagation?

7.8 A steel string of length 0.7 m has a diameter of 0.8 mm. What tensile stress must be set up in the string so that the fundamental frequency is 150 Hz? The density of steel is 7.8 Mg/m^3.

7.9 Derive the relationship between the transverse velocity of a particle of the string undergoing a standing wave and the longitudinal propagation speed of the transverse wave along the string.

7.10 A wire of 0.5 m length and 0.05 kg mass is stretched between a fixed end and a transverse harmonic vibrator. The tensile force is 10 N. If the distance between nodes along the wire is 15 cm and the peak-to-peak excursion at the antinodes is 5 cm, calculate the frequency and the amplitude of the transverse driving force. Use equation 7.17 in calculating the force.

7.11 Take equation 7.53 for the transverse vibrations of a solid bar and work out the simplifications for boundary conditions, one end free and the other end clamped.

7.12 A length of board 5 m long, 2 cm thick and 10 cm wide is held at the mid-point with the ends free. The fundamental period of free vibrations is 1 s. Take the density of the timber to be 900 kg/m³ and find the value for Young's modulus of the timber. If five such boards are clamped together, what will be the fundamental period of the bunch?

7.13 Derive the relationship between the propagation speed of transverse or flexural vibrations and that of longitudinal vibrations in a solid bar.

7.14 The speed of propagation of transverse waves along a solid rod is frequency dependent. If a pulse, that is, a band of frequencies, is travelling along the rod, the envelope of the so-called wave packet travels at the group speed defined thus $c_g = 1/(d\beta/d\omega)$. Find the formula for the group speed for such flexural waves in a rod.

7.15 An aluminium bar is 0.5 m long, 2 cm wide and 5 mm thick. Calculate the speeds of propagation of transverse and longitudinal waves of frequency 10 Hz in it. If the two ends are free what will be the two fundamental frequencies of free vibration, transverse and longitudinal, in the rod? The density of aluminium is 2.7 Mg/m³ and its Young's modulus is 0.07 GN/m².

7.16 Repeat problem 7.15 for a steel bar of identical dimensions. The density of steel is 7.8 Mg/m³ and its Young's modulus is 0.20 GN/m.

7.17 What are the frequencies of the lowest two overtones for transverse vibration in the aluminium and steel bars of problems 7.15 and 7.16?

7.18 A solid rod is clamped at one quarter of its length from one end and the ends are free. The speed of propagation of longitudinal waves in the rod is 5 km/s. The fundamental frequency is 3 kHz. What is the length of the rod?

7.19 The solid aluminium bar of problem 7.15 has the following boundary conditions: free at one end (at $x = 0$, $F_x = 0$) and, at $x = 0.5$ m, $F_x = 10 \cos \omega t$ N. What is the input or driving point impedance of the rod as a function of ω? What is the amplitude of the standing wave displacement?

7.20 A circular membrane is made of steel of thickness 0.1 mm and is stretched by a radial force per unit length of periphery of 10 kN/m. What radius must be chosen for a fundamental frequency of 200 Hz? What will the next two overtones be?

7.21 Repeat problem 7.20 for an aluminium membrane of the same thickness and tension.

7.22 A circular membrane of mass per unit area 1 kg/m^2 and diameter 1 m is required to have a fundamental frequency of 100 Hz. What tensile force per unit length of periphery must be applied?

7.23 A pipe of length 0.5 m is closed at one end. Its radius is 2 cm. Find the fundamental frequency and how many overtones of audible sound can be generated by it. Take the upper limit of audibility as 15 kHz.

7.24 What is the fundamental in the pipe of problem 7.23 and how many overtones are audible if the room temperature goes up from 20 °C to 30 °C?

7.25 What is the shortest length of a pipe of 5 cm radius (a) open at both ends and (b) closed at one end to resonate at the fundamental frequency of 500 Hz?

7.26 If an earth tremor travels 30 km in one minute and the wave is longitudinal what is the value of Young's modulus for the rocks in the path of the tremor? Take the density of the rocks to be 2.5 Mg/m^3.

Further Reading

Bishop, R.E.D., *Vibration* (Cambridge University Press, 1965)
Braddick, H.J.J., *Vibrations, Waves and Diffraction,* (McGraw-Hill, London, 1965)
Church, A.H., *Mechanical Vibrations,* 2nd edition (Wiley, New York, 1963)
Firth, I.M., Grant, D.F., and May, E.M., *Waves and Vibrations* (Penguin, London, 1973)
French, A.P., *Vibrations and Waves* (Nelson, London, 1971)
Haberman, C.M., *Vibration Analysis* (Merrill, Columbus, Ohio, 1968)
Kinsler, L.W. and Frey, A.R., *Fundamentals of Acoustics,* 2nd edition (Wiley, New York, 1962)
Main, I.G., *Vibrations and Waves in Physics* (Cambridge University Press, 1978)
Morse, P.M., *Vibration and Sound,* 2nd edition (McGraw-Hill, New York, 1948)
Sears, F.W. and Zemansky, M.W., *University Physics,* 4th edition (Addison-Wesley, Reading, MA., 1972)
Seto, W.W., *Theory and Problems of Mechanical Vibrations* (Schaum, New York, 1964)
Stephens, R.W.B. and Bate, A.E., *Acoustics and Vibrational Physics* (Arnold, London, 1966)
Stumpf, F.B., *Analytical Acoustics* (Ann Arbor Science, Ann Arbor, Mich., 1980)
Timoshenko, S., Young, D.H. and Weaver, W. Jr., *Vibration Problems in Engineering,* 4th edition (Wiley, New York, 1974)

8 The Piezoelectric Effect

8.1 Introduction and Objectives

The aim in this chapter is to define and describe the piezoelectric effect and to investigate the special features of longitudinal mechanical vibrations in piezoelectric crystals.

After reading this chapter the student should be able to

(a) define the direct and reverse aspects of the piezoelectric effect;
(b) derive the piezoelectric relationships between the mechanical quantities, stress and strain and the electrical quantities, field strength and dielectric displacement;
(c) derive a number of alternative pairs of independent simultaneous equations relating these four quantities;
(d) define the four piezoelectric constants and show the relationships between them;
(e) show how a mechanical longitudinal wave in a rod of a piezoelectric crystal can have electrical boundary conditions and initial conditions as well as mechanical;
(f) outline how such a rod can function as a source of mechanical vibrations;
(g) show how such a rod can perform as a stress transducer.

8.2 Basic Aspects of the Piezoelectric Effect

Most frequently piezoelectric crystals are encountered in the form of a short rod, a slab or disc as shown in figure 8.1 with the two opposite parallel faces coated with an electrically conducting metallic layer. The piezoelectric effect may be conveniently defined in relation to this geometry. This effect consists of two aspects

(a) The Direct Effect

If a stress is applied across the thickness of the crystal to produce a strain, a potential difference also develops between the electroded faces. As shown in figure 8.2, the potential difference is directly proportional to the stress whether the stress is negative (compressive) or positive (tensile).

THE PIEZOELECTRIC EFFECT

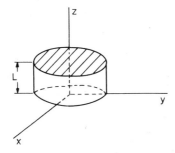

Figure 8.1 A cylindrical slab or disc of piezoelectric material of thickness L and the z-axis running along the main axis of the cylinder starting at the centre of one end. The two parallel faces are coated with electrically conducting electrodes.

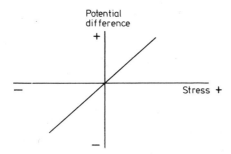

Figure 8.2 Application of a static stress across the piezoelectric cylinder produces a potential difference between the two electroded faces. The potential difference is directly proportional to the stress.

Furthermore, if the stress is alternating, the potential difference is also alternating (A.C.).

(b) The Reverse or Motor Effect

If a potential difference is applied across the crystal between the two electroded faces the crystal thickness is strained. This strain is directly proportional to the potential difference applied, as shown in figure 8.3. An alternating (A.C.) potential difference causes an alternating strain.

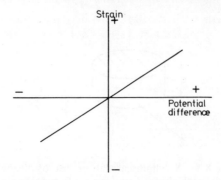

Figure 8.3 When a potential difference is applied between the two electroded faces of the piezoelectric crystal the cylinder is strained. The strain is directly proportional to the applied potential difference.

8.3 The Piezoelectric Relationships

Consider the slab of piezoelectric crystal shown in figure 8.1. The z-axis corresponds to the direction along the central axis of the crystal specimen. In this discussion variations in this direction only will be considered. Let T be the stress in this z-direction. This stress may vary with z but is constant for a given z value.

The stress within the piezoelectric material results in strain with the different planes of constant z within the crystal displaced to one extent or another from their equilibrium unstressed positions. If ζ is the displacement along the z-direction, the strain in this direction S is given by

$$S = \frac{\partial \zeta}{\partial z} . \tag{8.1}$$

If the displacement is small, the relationship between stress T and strain S is given by Hooke's law of equation 1.22 thus

$$T = cS \tag{8.2}$$

where c is the Young's modulus of elasticity, a constant for the material. This relationship can be expressed in the inverse form

$$S = \frac{1}{c} T$$

$$= \kappa T \tag{8.3}$$

where κ, the inverse of Young's modulus, is called the elastic compliance or compressibility of the piezoelectric substance.

THE PIEZOELECTRIC EFFECT

The materials that display piezoelectricity are, in general, electrical insulators. If an electric field strength or potential gradient E is established in the z-direction the resulting dielectric displacement D in the same direction is related linearly to E thus

$$D = \epsilon E \tag{8.4}$$

where ϵ is the dielectric constant of the material.

If the piezoelectric crystal is subjected to mechanical stress as well as electric field strength then the dielectric displacement D is a function of both of these parameters thus

$$D = \epsilon E + dT \; . \tag{8.5}$$

Here d is a piezoelectric constant for the material.

When the crystal is subjected to such a combination of mechanical stress and electrical field there is an increase in the internal energy U of the material. This increase in U per unit volume is

$$dU = T\, dS + E\, dD + \vartheta\, d\sigma \tag{8.6}$$

where ϑ is the absolute temperature and σ is the entropy. The increment in internal energy consists of stored mechanical energy, stored electrical energy and heat.

The increase in enthalpy per unit volume H is related to the change in internal energy thus

$$\begin{aligned} dH &= dU - d(TS) - d(ED) \\ &= -S\, dT - D\, dE + \vartheta\, d\sigma \; . \end{aligned} \tag{8.7}$$

Both dU and dH are exact differentials. This means that, as illustrated in figure 8.4 for dU, they depend only on the initial and final value and not on the path taken to achieve the change. Now from equation 8.6

$$T = \frac{\partial U}{\partial S} \tag{8.8}$$

while from equation 8.2

$$\begin{aligned} c &= \frac{\partial T}{\partial S} \\ &= \frac{\partial^2 U}{\partial S^2} \; . \end{aligned} \tag{8.9}$$

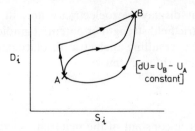

Figure 8.4 An exact differential dU where U is the internal energy does not depend on the manner or path of achieving the change in U but simply on the initial and final values of U. Other exact differentials are dH where H is enthalpy, and dG where G is Gibbs' free energy.

A third useful thermodynamic parameter is the Gibbs' free energy function G, defined thus

$$G = U - TS \tag{8.10}$$

from which one may find a third exact differential

$$dG = dU - T\,dS - S\,dT$$

$$= -S\,dT + E\,dD + \vartheta\,d\sigma \,. \tag{8.11}$$

Therefore

$$S = -\frac{\partial G}{\partial T} \tag{8.12}$$

and from equation 8.3

$$\kappa = \frac{\partial S}{\partial T}$$

$$= -\frac{\partial^2 G}{\partial T^2} \,. \tag{8.13}$$

From equation 8.7

$$S = -\frac{\partial H}{\partial T} \tag{8.14}$$

THE PIEZOELECTRIC EFFECT

and

$$D = -\frac{\partial H}{\partial E} . \tag{8.15}$$

From the general piezoelectric relationship of equation 8.5

$$d = \frac{\partial D}{\partial T}$$

$$= -\frac{\partial^2 H}{\partial T \partial E}$$

$$= \frac{\partial S}{\partial E} . \tag{8.16}$$

Therefore equation 8.3 may be expanded to apply to a piezoelectric substance

$$S = dE + \kappa T . \tag{8.17}$$

This equation, relating strain S to electric field strength E and mechanical stress T, and equation 8.5, relating dielectric displacement D to electric field strength E and mechanical stress T, are the two basic relationships describing the piezoelectric effect. Each of these equations couples together mechanical and electrical quantities in the piezoelectric material.

More formally precise versions of these two fundamental equations are

$$D = \epsilon^T E + dT \tag{8.18}$$

where ϵ^T is the dielectric constant at fixed stress T and

$$S = dE + \kappa^E t \tag{8.19}$$

where κ^E is the compliance of the material at fixed electric field strength E.

8.4 Alternative Forms of the Piezoelectric Relationships

Equations 8.18 and 8.9 relate the two parameters D and S respectively to the other two parameters E and T. Alternative interrelationships can be developed using simple algebra. It is of value to develop these alternatives in that they can be more convenient under different boundary conditions.

From equations 8.18 and 8.19

$$dT = D - \epsilon^T E$$
$$= D - \epsilon^T\left(\frac{S}{d} - \frac{\kappa^E}{d}T\right)$$

and so

$$T\left(d - \frac{\epsilon^T \kappa^E}{d}\right) = D - \frac{\epsilon^T}{d}S \ .$$

Therefore

$$T = -hD + c^D S \qquad (8.20)$$

where

$$h = \frac{d}{\epsilon^T \kappa^E - d^2} \qquad (8.21)$$

a new piezoelectric constant and

$$c^D = \frac{\epsilon^T}{\epsilon^T \kappa^E - d^2} \qquad (8.22)$$

the elastic modulus at constant dielectric displacement D.

Likewise, from the basic pair of simultaneous equations 8.18 and 8.19 can be written

$$D = \epsilon^T E + d\left(\frac{S}{\kappa^E} - \frac{d}{\kappa^E}E\right)$$
$$= \epsilon^S E - eS \qquad (8.23)$$

where

$$\epsilon^S = \frac{\epsilon^T \kappa^E - d^2}{\kappa^E} \qquad (8.24)$$

is the dielectric constant of the material at constant stress S and

$$e = \frac{d}{\kappa^E} \qquad (8.25)$$

is another piezoelectric constant of the crystal.

Likewise, the following can be derived from the basic equations

$$S = gD + \kappa^D T \qquad (8.26)$$

where

$$g = \frac{d}{\epsilon^T} \tag{8.27}$$

is a piezoelectric constant and

$$\kappa^D = \frac{\kappa^E \epsilon^T - d^2}{\epsilon^T} \tag{8.28}$$

is the compressibility of the material at constant D.

Alternative pairings that may be derived from equations 8.18 and 8.19 are

$$E = -hS + \frac{1}{\epsilon^S}D \tag{8.29}$$

$$E = -gT + \frac{1}{\epsilon^T}D \tag{8.30}$$

and

$$T = c^E S - eE \tag{8.31}$$

where

$$c^E = \frac{1}{\kappa^E} \tag{8.32}$$

is the elastic modulus at constant electric field strength E.

Equations 8.21, 8.22, 8.24, 8.25, 8.27, 8.28 and 8.32 detail a number of interrelationships between the various piezoelectric, mechanical and electrical constants of the piezoelectric material. Some of these relationships can be simplified if a further relationship is introduced thus

$$\frac{d^2}{\kappa^E \epsilon^T} = k^2 . \tag{8.33}$$

In that k couples together the piezoelectric constant d with the mechanical compliance constant κ^E and the dielectric constant ϵ^T, called the coupling coefficient of the material.

Equation 8.28 can be rewritten in terms of k, thus

$$\kappa^D = (1 - k^2)\kappa^E , \tag{8.34}$$

equation 8.24 becomes

$$\epsilon^S = (1 - k^2)\epsilon^T , \tag{8.35}$$

equation 8.21

$$h = \frac{1}{d}\left(\frac{k^2}{1-k^2}\right), \qquad (8.36)$$

equation 8.22

$$c^D = \frac{c^E}{(1-k^2)} \qquad (8.37)$$

and equations 8.25 and 8.27 yield

$$eg = k^2. \qquad (8.38)$$

Table 8.1 lists the values of some of the basic mechanical, dielectric and piezoelectric coefficients for a number of commonly used piezoelectric crystals, namely, quartz, barium titanate and two different formulations of lead zirconate titanate.

Table 8.1 Approximate Values of Basic Parameters of Common Piezoelectric Materials* (z Direction)

Material Parameter	Quartz (X-cut)	Barium titanate	Lead zirconate titanate (PZT-4)	Lead zirconate titanate (PZT-51)
ρ (10^3 kg/m^3)	2.7	5.5	7.5	7.7
$\varkappa^E$ (10^{-12} m^2/N)	11.4	5.6	13.5	17.8
ϵ^T (10^{-12} F/m)	42.0	1.2×10^3	15.5×10^3	24.8×10^3
d (10^{-12} C/N)	2.0	190.0	292.0	480.0
k^2	0.001	0.40	0.41	0.52

*Since both barium titanate and lead zirconate titanate are polycrystalline materials, they differ in formulation from one manufacturer to another and therefore also differ in precise values of parameters.

Note that each of the piezoelectric coefficients can have two different sets of units. Thus from equation 8.18 d may have units of D/T, (coulomb/square metre)/pascal or coulomb/newton, C/N. Equation 8.19 indicates that the units of d are those of S/E, metre/volt, m/V. Table 8.2 lists the pairs of sets of units corresponding to the four piezoelectric constants d, e, g and h.

THE PIEZOELECTRIC EFFECT

Table 8.2 Units of the Piezoelectric Constants

Constants	Units		
d	C/N	or	m/V
e	C/m²	or	N/V/m
g	m²/C	or	V/m/N
h	N/C	or	V/m

Example 8.1

Calculate the strains in discs of X-cut quartz and lead zirconate titanate-51 under conditions of constant dielectric displacement 50 C/m² and stress 5 Pa.

Solution

The relevant equation in this case is 8.26

$$S = gD + \kappa^D T$$

(a) X-cut Quartz

$$g = \frac{d}{\epsilon^T} = 0.048 \times 10^{-12} \text{ m}^2/\text{C} .$$

$$\kappa^D = (1 - k^2)\kappa^E \text{ from equation 8.34}$$

$$= 11.4 \times 10^{-12} \text{ m}^2/\text{N} .$$

Therefore

$$S = (2.4 + 57.0) \times 10^{-12}$$

$$= 59.4 \times 10^{-12} .$$

(b) Lead Zirconate Titanate–51

$$g = 0.02 \times 10^{-12} \text{ m}^2/\text{C and}$$

$$\kappa^D = 7.97 \times 10^{-12} \text{ m}^2/\text{N} .$$

Therefore

$$S = (1.0 + 39.9) \times 10^{-12}$$

$$= 40.9 \times 10^{-12} .$$

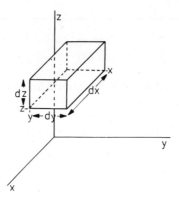

Figure 8.5 An elemental volume of the piezoelectric cylinder of figure 8.1. Located at x, y, z it has dimensions dx, dy and dz.

8.5 Wave Equation in a Piezoelectric Slab

Consider an elemental volume within the slab of piezoelectric crystal of figure 8.1 as illustrated in figure 8.5. As a result of the stress variations along the z-direction there is a net force F on the volume in the vertical direction where

$$F = (T_{z+dz} - T_z) \, dx \, dy . \tag{8.39}$$

Expanding T_{z+dz} in a Taylor series thus

$$T_{z+dz} = T_z + \frac{\partial T}{\partial z} dz \ldots$$

where the higher order terms may be neglected, the next equation may be developed from equation 8.39

$$F = \frac{\partial T}{\partial z} \, dx \, dy \, dz . \tag{8.40}$$

THE PIEZOELECTRIC EFFECT

This unbalanced force accelerates the element of volume in the vertical direction, thus, if ζ is displacement in this direction

$$\frac{\partial T}{\partial z} dx\, dy\, dz = \rho\, dx\, dy\, dz \ddot{\zeta}$$

or

$$\ddot{\zeta} = \frac{1}{\rho} \frac{\partial T}{\partial z} \tag{8.41}$$

where ρ is the density of the piezoelectric material. Equation 8.31, which relates the stress T to the strain S and the electric field strength E can be substituted for T in equation 8.41 to obtain

$$\ddot{\zeta} = \frac{1}{\rho}\left(c^E \frac{\partial S}{\partial z} - e\frac{\partial E}{\partial z}\right). \tag{8.42}$$

In most practical situations, the electric field strength is independent of z, that is, $\partial E/\partial z = 0$ and so from equations 8.41 and 8.1

$$\ddot{\zeta} = c^2 \frac{\partial^2 \zeta}{\partial z^2} \tag{8.43}$$

where

$$c = \frac{1}{\sqrt{(\rho \kappa^E)}}. \tag{8.44}$$

Equation 8.43 is a one-dimensional wave equation like equation 7.64 for the solid rod in longitudinal vibration. c is the speed of propagation. Equation 7.22 provides a model harmonic solution thus

$$\zeta = A_1 \sin(\omega t - \beta z) + B_1 \cos(\omega t - \beta z)$$
$$+ A_2 \sin(\omega t + \beta z) + B_2 \cos(\omega t + \beta z). \tag{8.45}$$

This solution consists of the sum of a forward and reverse-travelling wave.

Both electrical and mechanical boundary conditions and initial conditions are needed to determine the amplitude constants. Note firstly that the strain S at any axial position z and at any time t is given from equation 8.1 by

$$S = \frac{\partial \zeta}{\partial z}$$
$$= -\beta A_1 \cos(\omega t - \beta z) + \beta B_1 \sin(\omega t - \beta z)$$
$$+ \beta A_2 \cos(\omega t + \beta z) - \beta B_2 \sin(\omega t + \beta z). \tag{8.46}$$

Consider a number of sets of initial and boundary conditions as follows.

(a) No externally applied stress and electrical open-circuit between the two electrodes. This mechanical boundary condition means that at both $z = 0$ and $z = L$, $T = 0$. The electrical boundary condition entails no current flow into or out of the crystal at the electrodes. The current density $J = 0$ at $z = 0$ and at $z = L$. But

$$J = \frac{\partial D}{\partial t} . \tag{8.47}$$

Therefore at $z = 0$ and at $z = L$, D the dielectric displacement is either a constant or a function of z only.

Using equation 8.18, at $z = 0$, $E(0) = (1/\epsilon^T)D(0)$. Therefore from equations 8.19 and 8.46

$$S = \frac{d}{\epsilon^T} D(0)$$

$$= \beta(-A_1 + A_2) \cos \omega t + \beta(B_1 - B_2) \sin \omega t .$$

$D(0)$ is not a function of t and so must be equal to zero. Then

$$A_1 = A_2 = A$$

and

$$B_1 = B_2 = B .$$

At the other boundary $z = L$

$$S = -\beta A \cos(\omega t - \beta L) + \beta B \sin(\omega t - \beta L)$$

$$+ \beta A \cos(\omega t + \beta L) - \beta B \sin(\omega t + \beta L)$$

$$= -2\beta \sin \beta L (A \sin \omega t + B \cos \omega t)$$

$$= \frac{d}{\epsilon^T} D(TL) .$$

$D(L)$, however, is not a function of time t and so must be zero. For this condition to hold both A and B must also be zero and there is no wave, or

$$\sin \beta L = 0$$

that is

THE PIEZOELECTRIC EFFECT

$$\beta L = n\pi \text{ for } n = 1, 2, 3 \ldots$$

or

$$f = \frac{c\beta}{2\pi}$$
$$= n\frac{c}{2L} \text{ for } n = 1, 2, 3 \ldots$$

These are the frequencies of free vibratory waves in the crystal under these boundary conditions. The amplitude coefficients A and B must be found from the initial conditions of the vibration.

(b) No externally applied stress and an electrical short-circuit between the two electrodes. Therefore there is zero voltage difference between the electrodes and so $E(0) = E(L)$ for all t. Also at $z = 0$ and at $z = L$, $T = 0$.

At $z = 0$, from equation 8.19

$$S = dE(0)$$
$$= \beta(-A_1 + A_2) \cos \omega t + \beta(B_1 - B_2) \sin \omega t \ .$$

But $E(0)$ is fixed for all values of t and so must be zero. So

$$A_1 = A_2 = A$$

and

$$B_1 = B_2 = B \ .$$

Applying the boundary condition $T = 0$ at $z = L$ allows the conclusion that for free vibratory waves to exist in the crystal

$$\sin \beta L = 0$$

or

$$f = \frac{c\beta}{2\pi}$$
$$= n\frac{c}{2L} \text{ for } n = 1, 2, 3 \ldots$$

Once more these are the frequencies of the free vibratory waves sustainable in the crystal under these boundary conditions. The values of the amplitude coefficients A and B are set by the initial conditions.

(c) If at $z = 0$, $T = 0$ and $E = E_0 \cos \omega t$, while at $z = L$, $T = 0$ and $E = E_0 \cos \omega t$, there is no externally applied stress but there is an externally applied sinusoidal voltage difference between the two electrodes.

At $z = 0$, from equations 8.46 and 8.19

$$\beta(-A_1 + A_2) \cos \omega t + \beta(B_1 - B_2) \sin \omega t = dE_0 \cos \omega t .$$

Therefore

$$B_1 = B_2 = B$$

and

$$-A_1 + A_2 = \frac{d}{\beta} E_0 .$$

At $z = L$

$$\beta(-A_1 \cos \beta L + A_2 \cos \beta L - 2B \sin \beta L) \cos \omega t$$

$$- (A_1 \sin \beta L + A_2 \sin \beta L) \sin \omega t = dE_0 \cos \omega t$$

from which

$$A_1 = -A_2 = A$$

$$= -\frac{dE_0}{2\beta} .$$

Also

$$-2A \cos \beta L + 2\beta \sin \beta L = \frac{d}{\beta} E_0$$

$$= -2A ,$$

from which

$$B = -\frac{dE_0}{2\beta} \left(\frac{1 - \cos \beta L}{\sin \beta L} \right) .$$

The strain S at any point in the crystal may be shown to be

$$S = dE_0 \left[\cos \beta z + \left(\frac{1 - \cos \beta L}{\sin \beta L} \right) \sin \beta z \right] \cos \omega t .$$

The total strain S_T of the crystal slab may be derived

$$S_T = \frac{1}{L} \int_0^L S \, dz$$

$$= \frac{dE_0}{\beta L \sin \beta L}(\sin \beta L \sin \beta z - (1 - \cos \beta L) \cos \beta z \Big|_0^L)$$

$$= \frac{dE_0}{(\beta L/2)} \tan(\beta L/2).$$

Clearly resonances occur when

$$\frac{\beta L}{2} = \frac{(2n-1)\pi}{2} \qquad n = 1,2,3...$$

or

$$f = (2n-1)\frac{c}{2L}, \qquad n = 1,2,3...$$

Under these conditions the slab of piezoelectric material vibrates in the thickness mode. The frequency of the vibration is the frequency of the applied electrical voltage. Such an electrically driven crystal can serve to generate sound or ultrasound. High amplitude resonant vibrations can be achieved if the thickness of the slab is chosen such that

$$L = (2n-1)\frac{\lambda}{2}$$

an odd number of half-wavelengths.

(d) If at $z = 0$ and at $z = L$, $T = T_0 \cos \omega t$, and the two electrodes are open-circuited such that $D = 0$ at $z = 0$ and $z = L$.

In this case it can be shown that

$$B_1 = B_2$$

and

$$A_1 = -A_2$$
$$= -\frac{T_0}{2c^D \beta}.$$

The voltage drop across the crystal of amplitude V may be derived using equation 8.30

$$V = \int_0^L E\, dz$$
$$= -\frac{dT\,L}{\epsilon^T} \frac{\tan(\beta L/2)}{(\beta L/2)}$$

Figure 8.6 shows how the magnitude of this voltage varies with frequency.

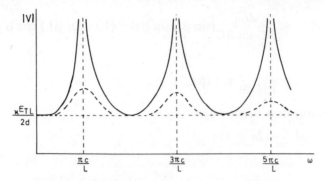

Figure 8.6 The absolute value of the amplitude of the potential difference developed across the piezoelectric transducer in open circuit which is subjected to an alternating stress of different frequencies $f [= \omega/(2\pi)]$. Resonances occur at $\omega = [(2n - 1)\pi c]/L$. The presence of damping (dashed curves) reduces the resonance heights.

At frequencies

$$f = (2n - 1)\frac{c}{2L}$$

such that

$$L = (2n - 1)\frac{\lambda}{2} \qquad n = 1,2,3...$$

voltage resonances occur. In the absence of damping these tend to infinity but damping limits them.

In this configuration, the crystal acts as an alternating pressure or stress transducer. The alternating output voltage has the same frequency as the stress and the amplitude of the voltage is proportional to the amplitude of the stress.

Example 8.2

What is the fundamental frequency for free vibration longitudinal standing waves in 1 cm lengths of barium titanate and lead zirconate titanate-4?

Solution

$$f_1 = \frac{c}{2L} = 0.5 \times c \times 10^2 \text{ Hz}.$$

For barium titanate from equation 8.44

$$c = \sqrt{[1/(5.5 \times 75.2 \times 10^{-9})]}$$

$$= 5.7 \text{ km/s} ,$$

and so

$$f_1 = 285 \text{ kHz} .$$

For lead zirconate titanate-4

$$c = \sqrt{[1/(7.5 \times 13.5 \times 10^{-9})]}$$

$$= 3.1 \text{ km/s}$$

and so

$$f_1 = 157 \text{ kHz} .$$

8.6 Experiments

Measurements on a Piezoelectric Disc

A convenient way of testing the direct piezoelectric effect is to apply a compressive stress to an electroded disc in a universal testing apparatus while at the same time measuring the voltage developed between the two faces. The voltage developed may be plotted as a function of stress. This may be done for a number of different piezoelectric materials. The universal tester can also measure the strain and so the piezoelectric relationship equation 8.19 can be checked for static conditions.

A D.C. voltage can also be applied to the disc and the resulting strain measured with the universal testing machine. If the voltage is maintained constant and then the stress is varied the compliance of the material at fixed electric field strength can be determined from equation 8.19. If on the other hand the stress is held constant and the applied voltage is varied, the same equation allows the piezoelectric constant d to be calculated.

Equation 8.18, the other piezoelectric relationship, can be tested in a similar way by applying both a stress and a voltage across the crystal while measuring the charge accumulated on it with an electrometer. D is the charge per unit area of the flat surface of the disc. By keeping the stress constant and varying the voltage, the slope of the D/E graph yields the dielectric constant at fixed stress. Conversely, keeping the voltage constant and varying the stress, the D/T graph has a slope equal to the piezoelectric constant d.

Similarly, experiments can be designed to test any other of the piezoelectric relationships using the same basic pieces of equipment.

Problems

8.1 Calculate the speed of longitudinal wave propagation in the four piezoelectric materials whose data are tabulated in table 8.1.

8.2 Calculate the half-wavelength thicknesses for each of the four materials at (a) 100 Hz, (b) 10 kHz and (c) 1 MHz.

8.3 Consider a disc of each of the four materials of 1 cm^2 area and 5 mm thickness.
(a) What is the open circuit voltage across each crystal when a compressive force of 100 N is applied across the thickness and zero dielectric displacement pertains?
(b) What is the net change in the thickness of each crystal when a potential difference of 500 V is impressed across the thickness from a very high impedance source and a stress of 10^5 N/m^2 exists in the thickness direction?

8.4 What is the fundamental frequency for longitudinal standing wave resonance for each of the four crystals of problem 8.3?

8.5 What compressive force must be applied across each crystal to generate a D.C. voltage of 100 V under conditions of zero dielectric displacement?

8.6 What voltage must be impressed across each crystal in order to cause a 1 per cent strain at atmospheric pressure of 10^5 Pa?

8.7 In the situation in problem 8.3(a) what charge should be taken from each crystal face to ensure that the voltage across each one is zero?

8.8 In the case of problem 8.3(b) what stress needs to be applied across each crystal in order to prevent any strain?

Further Reading

Beranek, L.L., *Noise and Vibration Control* (McGraw-Hill, New York, 1971)
Cady, W.G., *Piezoelectricity* (McGraw-Hill, New York, 1946)
Gooberman, G.L., *Ultrasonics, Theory and Application* (English Universities, London, 1968)
Kinsler, L.W. and Frey, A.R, *Fundamentals of Acoustics*, 2nd edition (Wiley, New York, 1962)
Mason, W.P., *Piezoelectric Crystals and Their Application to Ultrasonics* (Van Nostrand, Princeton, N.J., 1950)

9 Non-Linear Vibratory Systems

9.1 Introduction and Objectives

The aim in this chapter is to introduce the idea of mechanical elements, springs and dampers which do not behave linearly and the approximation approaches needed to solve the equations of motion of vibratory systems incorporating such elements.

After reading this chapter the student should be able to

(a) describe the characteristics of non-linear springs;
(b) outline the features of the free vibration of a mass/spring system in which the spring is non-linear;
(c) derive the approximate behaviour of this system when it is subjected to a harmonic forcing function;
(d) distinguish between viscous damping and hysteretic damping;
(e) describe the free vibration of a harmonic oscillator in which hysteretic damping occurs;
(f) outline the response of a harmonic oscillator which is hysteretically damped to a harmonic driving force;
(g) describe how static friction can affect the free oscillations of a system and cause a final error or offset.

9.2 Non-linear Spring

If any practical spring is stretched too much the force/deflection relationship becomes non-linear as indicated in figure 9.1. If the spring has an identical non-linear behaviour in tension and in compression, the force $F(x)$ is an odd function of the deflection x so that $F(-x) = -F(x)$. If a polynomial is used to represent $F(x)$, to a first approximation only the first non-linear term, the cubic term, needs to be included thus

$$F(x) = Cx + \delta Cx^3$$
$$= (1 + \delta x^2) Cx \qquad (9.1)$$

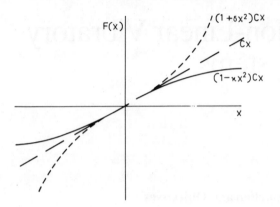

Figure 9.1 The two types of non-linear spring deflection characteristics which have identical behaviour in tension and compression. One such spring has a constant that rises with deflection — a stiffening spring. The other, the softening spring, has a constant which falls with increasing deflection.

or

$$F(x) = Cx - \kappa C x^3$$

$$= (1 - \kappa x^2)Cx \qquad (9.2)$$

where C is the Hooke's law spring constant of equation 1.24 and δ and κ are constants. For the linear case δ and κ are zero. The spring characterised by equation 9.1 is a stiffening spring, the stiffness or slope increasing with the deflection. Conversely, the spring of equation 9.2 has a stiffness falling with increasing deflection and is termed a softening spring.

The polynomial chosen to represent the force function $F(x)$ has odd powers only in order to represent the odd function $F(x)$. If the spring force/deflection relationship is not symmetrical in tension and compression, a quadratic or higher order function would have to be chosen.

Consider the simple mass/spring oscillator horizontally mounted as in figure 9.2 under free vibrations. Assume the spring to have a characteristic function as given in equation 9.1 in which $\delta x^2 \ll 1$, that is, slight non-linearity. The equation of motion may be written following equation 2.3

$$\ddot{x} + (1 + \delta x^2)\omega_0^2 x = 0 \; . \qquad (9.3)$$

NON-LINEAR VIBRATORY SYSTEMS

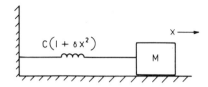

Figure 9.2 A simple harmonic oscillator with a stiffening non-linear spring. The single degree of freedom is along the x-axis and the origin is at the resting position of the mass.

A number of features of the resulting free vibrations may be deduced from the features of the system and by reference to the simple harmonic oscillator described in chapter 2.

(a) The motion of the mass is periodic with period T such that the displacement $x(t)$ must display the following relationship

$$x(t) = x(t + T) \ . \tag{9.4}$$

(b) The oddness of the spring force as a function of displacement demands that the negative phase of the displacement period be a mirror-image about $x = 0$ of the positive phase with each of these phases therefore lasting for $T/2$ s. Consequently

$$x\left(t + \frac{T}{2}\right) = -x(t) \ . \tag{9.5}$$

For a periodic solution, the Fourier theory of harmonic analysis guides one to search for some combination of harmonics. However, due to condition 9.5 which rests on the symmetry of the spring force, even harmonics are excluded. The fundamental frequency will be $f_f \ (= 1/T)$ so that $\omega_f = 2\pi/T$. If the system is initially at rest, $\dot{x}(0) = 0$, all of the phase angles (see equation 2.10) will be zero and the trial solution can be written

$$x = A_1 \cos \omega_f t + A_3 \cos 3\omega_f t + \ldots \tag{9.6}$$

a sum of the fundamental and higher odd harmonics. Take the ratio of the third harmonic amplitude to that of the fundamental as s where

$$s = \frac{A_3}{A_1} \ . \tag{9.7}$$

Since the degree of non-linearity in the spring is small it is reasonable to assume that the extent of the departure of the solution from the fundamental is small, that is, $s \ll 1$. Therefore s^2, s^3 and higher powers of s are negligible. Now substitute equation 9.6 into 9.3 in order to find ω_f and s, but first work out the individual terms in this latter equation

$$\ddot{x} = -\omega_f^2 A_1 (\cos \omega_f t + 9s \cos 3\omega_f t \ldots)$$

and

$$x^3 = A_1^3 (\cos^3 \omega_f t + 3s \cos^2 \omega_f t \cos 3\omega_f t \ldots) \ .$$

Therefore from equation 9.3

$$-\omega_f^2 A_1 (\cos \omega_f t + 9s \cos 3\omega_f t) + \omega_0^2 A_1 (\cos \omega_f t + s \cos 3\omega_f t)$$
$$+ \delta \, \omega_0^2 A_1^3 (\cos^3 \omega_f t + 3 s \cos^2 \omega_f t \cos 3\omega_f t) = 0 \ . \tag{9.8}$$

Since $\delta A_1^2 \ll 1$, $\delta A_1^2 s$ is negligible. Also

$$\cos^3 \omega_f t = \tfrac{1}{4} \cos 3\omega_f t + \tfrac{3}{4} \cos \omega_f t \ .$$

Therefore equation 9.8 becomes

$$\cos \omega_f t \, (-\omega_f^2 + \omega_0^2 + \tfrac{3}{4}\delta \omega_0^2 A_1^2)$$
$$+ \cos 3\omega_f t \, (-9 s\omega_f^2 + s\omega_0^2 + \tfrac{1}{4}\delta \omega_0^2 A_1^2) = 0 \ . \tag{9.9}$$

This equation holds for all values of t, however, and so the separate coefficients of $\cos \omega_f t$ and of $\cos 3\omega_f t$ must equal zero. That is

$$-\omega_f^2 + \omega_0^2 + \tfrac{3}{4}\delta \omega_0^2 A_1^2 = 0$$

or

$$\omega_f^2 = \omega_0^2 (1 + \tfrac{3}{4}\delta A_1^2) \tag{9.10}$$

and

$$-9s\omega_f^2 + s\omega_0^2 + \tfrac{1}{4}\delta \omega_0^2 A_1^2 = 0 \ . \tag{9.11}$$

Substituting equation 9.10 into 9.11, yields

$$-9s\omega_0^2 - \frac{27}{4} s\omega_0^2 \delta A_1^2 + s\omega_0^2 + \tfrac{1}{4}\delta \omega_0^2 A_1^2 = 0$$

from which

$$s\left(8 + \frac{27}{4}\delta A_1^2\right) = \frac{1}{4}\delta A_1^2 .$$

But since $\delta A_1^2 \ll 1$

$$s = \frac{1}{32}\delta A_1^2 . \tag{9.12}$$

Thus the departure from the pure harmonic in the solution is directly proportional to the non-linearity of the spring (δ) and becomes greater at higher amplitudes as the square of the amplitude. This amplitude A_1 is determined by the initial conditions such as the value of $x(0)$.

From equation 9.10 the period T is found

$$T = \frac{2\pi}{\omega_f}$$

$$= \frac{2\pi}{\omega_0 \sqrt{(1 + \tfrac{3}{4}\delta A_1^2)}} . \tag{9.13}$$

Note that the hardening spring decreases the fundamental period from that found with a Hookean spring in equation 2.13. A softening spring produces an increase in the fundamental period. It is also noteworthy that the initial conditions which set the value of A_1 thereby affect both the departure from pure harmonic behaviour (s) and the fundamental period of free vibrations (T). By contrast, in the simple harmonic oscillator described in chapter 2 pure harmonic vibrations only can occur and the period is determined by the system parameters C and M exclusively.

In summary, the main effect of the non-linearity of the spring is to introduce higher harmonics into the free vibratory motion of the mass.

Example 9.1

A mass of 10 kg is attached to a spring as in figure 9.2. When the initial displacement is 10^{-2} m, the period of the oscillations is 1 s but, when the initial displacement is 10^{-1} m, the period is 0.98 s. Write down the expression for the non-linear spring constant.

Solution

$$T_0 = 1 = 2\pi\sqrt{(M/C)} = 2\pi/\omega_0 .$$

Therefore

$$C = 4\pi^2 M = 394.8 \text{ N/m} .$$

From equation 9.13

$$T_2 = \frac{T_0}{\sqrt{(1 + \tfrac{3}{4}\delta A_1^2)}} = 0.98 \text{ and so}$$

$$1.04 = 1 + \tfrac{3}{4}\delta A_1^2$$

$$\delta A_1^2 = 0.05.$$

Since $s \ll 1$, take $A_1 = 10^{-1}$ as an approximation. Then $\delta = 5$. So from equation 9.1 the non-linear spring constant is $394.8(1 + 5x^2)$ N/m.

9.3 Non-linear Spring and Mass under Harmonic External Force

Consider the system of figure 9.2 in which the mass is subjected to an external driving force $f(t) = F_0 \cos \omega t$ acting along the x-axis. Now the equation of motion of the system is

$$\ddot{x} + \omega_0^2(1 + \delta x^2)x = \frac{F_0}{M}\cos \omega t. \tag{9.14}$$

This equation is called the Duffing equation and its general steady-state solution has not been worked out. It is possible, however, to develop an approximation to the solution. Assume as before only slight non-linearity such that the solution is close to harmonic thus

$$x_s = A \cos \omega t. \tag{9.15}$$

This assumed solution will satisfy the equation of motion 9.14 only at $t = \pi/2\omega, 3\pi/2\omega \ldots (2n-1)\pi/2\omega$, $(n = 1, 2, 3 \ldots)$ since both $f(t)$ and x_s are zero then but it will not satisfy the equation of motion at any other times. It can be made to satisfy it also at $t = \pi/\omega, 2\pi/\omega \ldots n\pi/\omega$, $(n = 1, 2, 3 \ldots)$ by judicious choice of A. At these latter instants $x_s = A$ and $f(t) = F_0$ so equation 9.14 becomes

$$-\omega^2 A + \omega_0^2 A + \omega_0^2 \delta^2 A^3 = \frac{F_0}{M}$$

or

$$\left(\frac{F_0}{M} + \omega^2 A\right) = \omega_0^2 A + \omega_0^2 \delta^2 A^3. \tag{9.16}$$

This equation may be solved graphically as in figure 9.3. The right-hand side and the left-hand side are plotted separately against A and the points of intersection between the two graphs give the solutions. Since the left-hand side is variable because ω is variable, a range of solutions is possible.

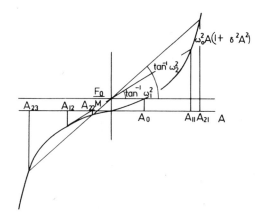

Figure 9.3 The graphical approach to solving equation 9.16. Both $[(F_0/M) + \omega^2 A]$ and $(\omega_0^2 A + \omega_0^2 \delta^2 A^3)$ are plotted against A and the values of A at the intersections between these curves are the solutions.

When $\omega = 0$, $A = A_0$. Between $\omega = 0$ and $\omega = \omega_1$ where $(F_0/M + A\omega_1^2)$ becomes tangential to $(\omega_0^2 A + \omega_0^2 \delta^2 A^3)$ on the negative side there is but one value of A at each ω value. At $\omega = \omega_1$ there are two intersections giving two values of A, A_{11} and A_{12}. At $\omega > \omega_1$, for instance at ω_2, there are three intersections A_{21} (positive) and A_{22} and A_{23} (both negative). The solutions A are plotted against ω in figure 9.4.

At high frequencies three amplitudes are possible, two of them out of phase by π rad with the forcing function. At $\omega = \omega_1$, two amplitudes are possible, one in phase and one π rad out of phase with the driving force. At low frequencies only one amplitude is allowed and the displacement is then in phase with the force.

If the frequency is such that the operating amplitude is that of point D in figure 9.4 and the frequency is reduced the amplitude will increase negatively following the operating point along the curve DFE to E. Then it will jump to Q without a change in frequency. Further reduction in frequency causes the operating point to move along QR, the amplitude falling all the way.

If on the other hand the frequency is raised from an operating point near R, the amplitude will proceed to infinity along the operating curve RQP causing catastrophic failure. If there is enough damping present, the amplitude may jump from P to F with no change in frequency. Such jumping behaviour or instability is characteristic of non-linear oscillating systems.

An alternative approximate approach, as follows, gives another insight

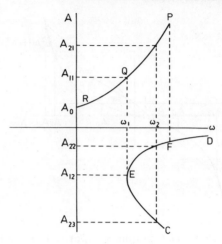

Figure 9.4 The solutions to equation 9.16, as obtained in figure 9.3, depend on ω. As ω is increased, different sets of solutions are found. The solutions A, are plotted against $\omega (= 2\pi f)$ where f is the frequency of the driving force. In the range $0 \leq \omega \leq \omega_1$, only one solution A exists. At $\omega = \omega_1$, two solutions A_{11} and A_{12} exist. At $\omega > \omega_1$, three solutions A_{21}, A_{22} and A_{23} exist.

into this non-linear system. Substitute $x_s = A \cos \omega t$ into equation 9.14 but do not substitute for $\ddot{x}$ thus

$$\ddot{x}_s + \omega_0^2 A \cos \omega t + \omega_0^2 \delta A^3 \cos^3 \omega t = \frac{F_0}{M} \cos \omega t$$

or

$$\ddot{x}_s = \left(\frac{F_0}{M} - \omega_0^2 A - \tfrac{3}{4}\omega_0^2 \delta A^3\right) \cos \omega t - \tfrac{1}{4}\omega_0^2 \delta A^3 \cos 3\omega t. \quad (9.17)$$

Therefore integrating each side of the equation twice

$$\dot{x}_s = \frac{1}{\omega}\left(\frac{F_0}{M} - \omega_0^2 A - \tfrac{3}{4}\omega_0^2 \delta A^3\right) \sin \omega t - \frac{1}{12\omega}\omega_0^2 \delta A^3 \sin 3\omega t$$

and

$$x_s = -\frac{1}{\omega^2}\left(\frac{F_0}{M} - \omega_0^2 A - \tfrac{3}{4}\omega_0^2 \delta A^3\right) \cos \omega t + \frac{1}{36\omega^2}\omega_0^2 \delta A^3 \cos 3\omega t \ .$$

(9.18)

This approximation to the steady state response thus contains one component at the fundamental driving frequency but also a third harmonic

NON-LINEAR VIBRATORY SYSTEMS

component. Once more in this case the generation of higher harmonics is an indicator of non-linearity.

If the driving force function has within it a frequency equal to three times the system resonant frequency, then a resonance will build up including both the system third harmonic and the fundamental. This phenomenon is called sub-harmonic generation and it occurs only in non-linear systems.

Example 9.2

If the mass in example 9.1 experiences a force 10 cos 4t N and the maximum observed displacement is 0.11 m find the amplitudes of the first and third harmonics of the motion of the mass.

Solution

From equation 9.18 the displacement of the mass is given as

$$x_s = A_1 \cos 4t + A_3 \cos 12t$$

where $A_1 + A_3 = 0.11$ and

$$A_1 = \frac{1}{\omega^2}\left(\omega_0^2 A + \frac{3}{4}\omega_0^2 \delta A^3 - \frac{F_0}{M}\right)$$

while

$$A_3 = \frac{1}{\omega^2}\left(\frac{\omega_0^2 \delta A^3}{36}\right).$$

Using the parameter values from example 9.1

$$39.4 A + 147.9 A^3 - 1 + 5.5 A^3 = 1.76$$

$$A(39.4 + 153.4 A^2) = 2.6 \ .$$

By trial and error $A = 0.07$ m and so

$$A_1 = 0.11 \text{ m and } A_3 = 0.0001 \text{ m} \ .$$

9.4 Non-linear Damping in a Freely Vibrating Simple Oscillator

The rate of energy loss in a system with a viscous damping is predicted from equation 3.45 to rise with the square of the frequency of the oscillation. In many systems where the damping arises from internal mechanisms within solids, the energy loss in practice tends to remain constant or even to fall with rising frequency. If harmonic motion is involved, the damping coefficient is still proportional to relative velocity but also inversely proportional to frequency. Such hysteretic damping force may be represented, again assuming harmonic behaviour

$$F_d = -\frac{h}{\omega}\dot{x} \tag{9.19}$$

where h is the hysteretic damping coefficient. The units of h are newtons/metre (N/m). Schematically such a damper is represented by a crossed-box symbol as in figure 9.5.

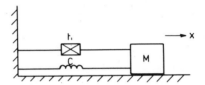

Figure 9.5 A damped harmonic oscillator in which the damper is non-linear of the hysteretic type where h is the hysteretic damping coefficient.

The equation of motion of the oscillator of figure 9.5 is, after equation 3.17

$$\ddot{x} + \frac{h}{M\omega}\dot{x} + \omega_0^2 x = 0 \ . \tag{9.20}$$

Assume as one initial condition $\dot{x}(0) = 0$ and try as a solution

$$x = A\, e^{-\alpha t} \cos \omega_f t \tag{9.21}$$

In order to find α and ω_f substitute this value of x into equation 9.20 but firstly note that

$$\omega_0^2 x = \omega_0^2 A\, e^{-\alpha t} \cos \omega_f t$$

$$\frac{h}{M\omega_f}\dot{x} = -\frac{h\alpha}{M\omega_f} A\, e^{-\alpha t} \cos \omega_f t - \frac{h}{M} A\, e^{-\alpha t} \sin \omega_f t$$

and

$$\ddot{x} = (\alpha^2 - \omega_f^2) A\, e^{-\alpha t} \cos \omega_f t + 2\alpha\omega_f A\, e^{-\alpha t} \sin \omega_f t \ .$$

NON-LINEAR VIBRATORY SYSTEMS

Therefore equation 9.20 becomes

$$0 = \left(\alpha^2 - \omega_f^2 + \omega_0^2 - \frac{h\alpha}{M\omega_f}\right) A\, e^{-\alpha t} \cos \omega_f t$$

$$+ \left(2\alpha\omega_f - \frac{h}{M}\right) A\, e^{-\alpha t} \sin \omega_f t$$

or

$$\left(\alpha^2 - \omega_f^2 + \omega_0^2 - \frac{h\alpha}{M\omega_f}\right) \cos \omega_f t + \left(2\alpha\omega_f - \frac{h}{M}\right) \sin \omega_f t = 0.$$

(9.22)

This equation holds at all values of t and so each amplitude term must be individually equal to zero. Therefore

$$2\alpha\omega_f - \frac{h}{M} = 0$$

or

$$\alpha = \frac{h}{2M\omega_f}. \tag{9.23}$$

Also

$$\alpha^2 - \omega_f^2 - \omega_0^2 - \frac{h\alpha}{M\omega_f} = 0$$

which by substituting for α from equation 9.23 simplifies to

$$\omega_f^4 - \omega_0^2\omega_f^2 + \frac{h^2}{4M^2} = 0. \tag{9.24}$$

The solutions to this quadratic equation in ω_f^2 may be written

$$\omega_f^2 = \frac{\omega_0^2}{2} \pm \tfrac{1}{2}\sqrt{\left(\omega_0^4 - \frac{h^2}{M^2}\right)}. \tag{9.25}$$

Since ω_f^2 must be positive, both of the above solutions may be taken as valid thus

$$\omega_{f1}^2 = \tfrac{1}{2}\omega_0^2 + \tfrac{1}{2}\omega_0^2\sqrt{\left(1 - \frac{h^2}{M^2\omega_0^4}\right)}$$

and

$$\omega_{f2}^2 = \tfrac{1}{2}\omega_0^2 - \tfrac{1}{2}\omega_0^2\sqrt{\left(1 - \frac{h^2}{M^2\omega_0^4}\right)}. \tag{9.26}$$

Here $\omega_{f1} > \omega_{f2}$. The corresponding values of α are given from equation 9.23

$$\alpha_1 = \frac{h}{2M\omega_{f1}}$$

and

$$\alpha_2 = \frac{h}{2M\omega_{f2}} \cdot \qquad (9.27)$$

Note that $\alpha_2 > \alpha_1$ so that the higher frequency component in the solution is more slowly damped than the lower frequency component. This is of the nature of hysteretic damping.

In summary, the free vibration consists of two damped sinusoids

$$x = \omega_0^2 A_1 \, e^{-\alpha_1 t} \cos \omega_{f1} t + \omega_0^2 A_2 \, e^{-\alpha_2 t} \cos \omega_{f2} t \qquad (9.28)$$

Example 9.3

A mechanical vibrator with a mass of 5 kg and spring constant 200 N/m has hysteretic damping the coefficient of which is to be determined. The mass is given an initial displacement of 2 cm and released from rest. After 200 s the amplitude of the displacement is 0.1 mm and the period is 0.95 s.

Solution

Here $\omega_0^2 = 40$. Since $x(0) = 0.02$ m

$$A_1 + A_2 = 0.5 \text{ mm}$$

Assume that after 200 s the lower frequency component is no longer appreciable. Then from equation 9.26, since

$$\omega_{f1} = 6 \text{ rad/s},$$

$$h^2 = 0.2 \, M^2 \omega_0^4, \qquad \text{and}$$

$$h \,. = 89.4 \text{ N/m}.$$

But is the assumption valid? From equation 9.26

$$\omega_{f2} = 1.48 \text{ rad/s}$$

and so

NON-LINEAR VIBRATORY SYSTEMS

$\alpha_1 = 6.04$ 1/s and $\alpha_2 = 1.49$ 1/s from equation 9.27 .

Since the decay constant for the lower frequency component is some four times that of the higher frequency element, it is not too unreasonable to take the displacement after an interval of some 200 periods to constitute the higher frequency element only.

9.5 Oscillator with Hysteretic Damping under Harmonic Force

If the system of figure 9.5 is subjected to a force $f(t) = F_0 \cos \omega t$, the equation of motion becomes after equation 4.2,

$$\ddot{x} + \frac{h}{M\omega}\dot{x} + \omega_0^2 x = \frac{F_0}{M} \cos \omega t . \tag{9.29}$$

Take as a trial solution in the steady-state

$$x = A \cos \omega t + B \sin \omega t \tag{9.30}$$

and by substitution in equation 9.29

$$A = \frac{(F_0/M)(\omega_0^2 - \omega^2)}{(\omega_0^2 - \omega^2)^2 + (h/M)^2} \tag{9.31}$$

and

$$B = \frac{F_0 h/M^2}{(\omega_0^2 - \omega^2)^2 + (h/M)^2} . \tag{9.32}$$

These two coefficients vary with $\omega(= 2\pi f)$ where f is the driving frequency as shown in figure 9.6. Clearly a displacement resonance occurs.

To study the power resonance, the average rate of work done in overcoming the hysteretic damping is the average power into the system thus

$$P = \frac{1}{T} \int_t^{t+T} \frac{h}{\omega}(\dot{x})^2 \, dt$$

$$= \frac{h}{T\omega} \int_t^{t+T} (\omega^2 A^2 \sin^2 \omega t + \omega^2 B^2 \cos^2 \omega t - 2\omega^2 AB \sin \omega t \cos \omega t) \, dt$$

$$= \frac{h}{2\omega}(\omega^2 A^2 + \omega^2 B^2)$$

$$= \frac{hF_0^2}{2M^2}\left(\frac{\omega}{(\omega_0^2 - \omega^2)^2 + (h/M)^2}\right) . \tag{9.33}$$

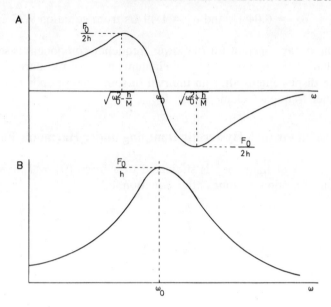

Figure 9.6 The amplitudes A and B, of the cosine and sine terms respectively, of the steady-state response of the hysteretically damped harmonic oscillator, to a harmonic driving force of different frequencies f ($= \omega/2\pi$). A goes to a peak at $\omega < \omega_0$, is zero at $\omega = \omega_0$ and goes to a negative peak at $\omega > \omega_0$. B is positive for all values of ω and is a maximum at $\omega = \omega_0$.

By differentiating the term within the brackets with relation to ω and setting the result equal to zero, it can be shown that the peak of the average power P does not occur at $\omega = \omega_0$ but at a higher frequency $f_m (= \omega_m/2\pi)$, such that

$$\omega_m^2 = \tfrac{1}{3}\omega_0^2 + \tfrac{2}{3}\omega_0^2 \sqrt{[1 + (3h^2/4M^2\omega_0^2)]} \ . \tag{9.34}$$

The power resonance curve therefore looks like that of figure 9.7 and it is not symmetrical about $\omega = \omega_m$. For instance the half-power points may be graphically determined as follows. Firstly set

$$\frac{\omega}{(\omega_0^2-\omega^2)^2 + h^2/M^2} = \tfrac{1}{2}$$

or

$$2\omega - h^2/M^2 = (\omega_0^2 - \omega^2)^2 \ . \tag{9.35}$$

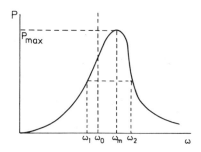

Figure 9.7 The power resonance curve for the driven oscillator with hysteretic damping. The power dissipation is a maximum at $\omega_m > \omega_0$. The curve is not symmetrical.

This equation may be solved graphically using figure 9.8 which contains plots of $(2\omega - h^2/M^2)$ and of $(\omega_0^2 - \omega^2)^2$ against ω. The intersections of these two graphs at ω_1 and ω_2 give the half-power frequencies f_1 and f_2.

In summary, this type of damping in an oscillator causes resonant behaviour somewhat different from that found in the viscous damping case. The resonant frequency f_m is greater than f_0 which would be the free vibration frequency for viscous damping. f_m is also greater than either of the two possible free vibration frequencies, f_{f1} or f_{f2}, both of which (see equation 9.26) are less than f_0 in this system. Furthermore, in general, the power resonance curve is not symmetrical and the half-power points must be determined graphically.

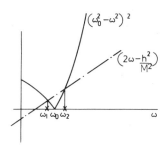

Figure 9.8 A graphical means is used to solve equation 9.35 and to find the half power values of ω, ω_1 and ω_2. Thus $[2\omega - (h^2/M^2)]$ and $(\omega_0^2 - \omega^2)^2$ are plotted against ω and the points of intersection yield the values of ω_1 and ω_2.

Example 9.4

Find the power resonance frequency for the hysteretically damped vibrator of example 9.3.

Solution

In this system h = 89.4 N/m and $\omega_0^2 = 40$ and so from equation 9.34

$$\omega_m^2 = 13.3 + 26.7 \times 2.64$$

$$\omega_m = 6.53 \text{ rad/s} .$$

So

$$f_m = 1.04 \text{ Hz} .$$

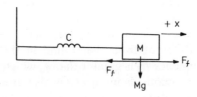

Figure 9.9 A simple harmonic oscillator in which static friction applies between the moving mass and the supporting table.

9.6 Static Friction Damping in an Oscillator

Occasionally the damping due to friction, even in a dynamic oscillating system, can best be approximated by the static or limiting friction mechanism. In this case the frictional force is proportional to the normal force between the mass and the supporting horizontal table

$$F_f = \mu M g$$

(equation 3.1). This force is independent of displacement, velocity or frequency but it opposes the motion in either direction.

Thus for positive $\dot{x}$ motion in figure 9.9 the equation of motion is

$$\ddot{x} + \mu g + \omega_0^2 x = 0 \tag{9.36}$$

NON-LINEAR VIBRATORY SYSTEMS

and the solution is

$$x = A_1 \cos \omega_0 t + B_1 \sin \omega_0 t - \frac{\mu g}{\omega_0^2} . \tag{9.37}$$

For negative $\dot{x}$ motion, the equation of motion becomes

$$\ddot{x} - \mu g + \omega_0^2 x = 0 \tag{9.38}$$

and the solution is

$$x = A_2 \cos \omega_0 t + B_2 \sin \omega_0 t + \frac{\mu g}{\omega_0^2} . \tag{9.39}$$

The amplitude constants A_1, A_2, B_1 and B_2 depend on the initial conditions. As an example, take as initial conditions $x(0) = X_0$ and $\dot{x}(0) = 0$ with the motion starting in the negative direction and the velocity $\dot{x}$ negative. Then from equation 9.39

$$A_2 = X_0 - \frac{\mu g}{\omega_0^2} . \tag{9.40}$$

Also

$$0 = \omega_0 B_2 \cos \omega_0 t$$

or

$$B_2 = 0 . \tag{9.41}$$

Therefore equation 9.39 becomes

$$x = \left(X_0 - \frac{\mu g}{\omega_0^2} \right) \cos \omega_0 t + \frac{\mu g}{\omega_0^2} . \tag{9.42}$$

This solution applies only when the mass is moving from right to left. Therefore at $t = \pi/\omega_0$ the direction of motion is reversed, the velocity becomes positive and the basic equation of motion is then equation 9.36. At $t = \pi/\omega_0$, however, the new initial conditions for this equation are

$$x\left(\frac{\pi}{\omega_0}\right) = -\left(X_0 - \frac{2\mu g}{\omega_0^2} \right)$$

and

$$\dot{x}\left(\frac{\pi}{\omega_0}\right) = 0 .$$

Therefore, from the solution of equation 9.37

$$-X_0 + \frac{2\mu g}{\omega_0^2} = A_1 - \frac{\mu g}{\omega_0^2}$$

or

$$A_1 = -\left(X_0 - \frac{3\mu g}{\omega_0^2}\right) \tag{9.43}$$

and

$$B_1 = 0 . \tag{9.44}$$

Equation 9.37 then becomes

$$x = -\left(X_0 - \frac{3\mu g}{\omega_0^2}\right) \cos \omega_0 t - \frac{\mu g}{\omega_0^2} . \tag{9.45}$$

Proceeding in this fashion, it can be shown that the third half-cycle of the motion is described by

$$x = \left(X_0 + \frac{5\mu g}{\omega_0^2}\right) \cos \omega_0 t + \frac{\mu g}{\omega_0^2} \tag{9.46}$$

and the fourth half-cycle is given by

$$x = -\left(X_0 - \frac{7\mu g}{\omega_0^2}\right) \cos \omega_0 t - \frac{\mu g}{\omega_0^2} \tag{9.47}$$

and so on.

The decrement, or reduction in amplitude per period, is linear and is given by

$$X_{n-1} - X_n = \frac{4\mu g}{\omega_0^2} . \tag{9.48}$$

Figure 9.10 illustrates the free oscillation of this system. The envelope is a straight line and not an exponential decay as in the case of viscous damping. The frequency is the undamped natural frequency.

When the magnitude of x becomes less than $\mu g/\omega_0^2$, motion stops with x in the range $+ \mu g/\omega_0^2$ to $- \mu g/\omega_0^2$. The mass comes to rest at an offset or error, to the right or left of the location where the spring is at its resting length, that is, with the spring under strain and exerting a force of up to $\pm \mu M g$.

Example 9.5

An oscillator such as that of figure 9.9 is found to vibrate freely with period 2 s with one excursion of the mass smaller by 0.9 than the previous one of

NON-LINEAR VIBRATORY SYSTEMS

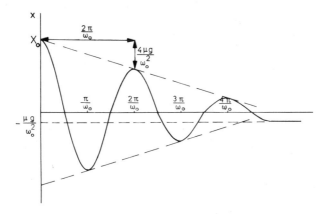

Figure 9.10 The free-vibration displacement of the mass in the vibrator with static friction damping. It is a decaying free oscillation of period $2\pi/\omega_0$. The decay is linear and amounts to $4\mu g/\omega_0^2$ each period. There is a final off-set of $\pm\ \mu g/\omega_0^2$.

10 cm. Find the final resting position of the 10 kg mass, the spring constant and the coefficient of static friction.

Solution

Here $\omega_0 = 3.14$ rad/s and $\omega_0^2 = 9.87$. Taking the acceleration due to gravity $g = 9.8$ m/s², from equation 9.48 the coefficient of friction is found

$$\mu = 0.01 \times \omega_0^2/(4 \times g) = 0.0025 \ .$$

$$C = \omega_0^2 \times M = 98.7 \text{ N/m} \ .$$

The final position of the mass will, from figure 9.10, be $\pm\ 0.0025$ m.

9.7 Experiments

(a) Non-linear Simple Pendulum

The basic formulae for the simple pendulum hold for very small angular deflection, but the true period is a function of the maximum angular deflection ϕ, as well as of L and g, thus

$$T = 2\pi \sqrt{\left(\frac{L}{g}\right)} \left(1 + \frac{1^2}{2^2} \sin^2 \frac{\phi}{2} + \frac{1^2 \times 3^2}{2^2 \times 4^2} \sin^4 \frac{\phi}{2} + \ldots\right).$$

The period of the simple pendulum may be measured for a range of values of ϕ and this relationship checked. T^2 may be plotted versus ϕ and the value of ϕ determined at which the departure from the constant level becomes appreciable.

(b) A Non-linear Spring System

A mass with a horizontal degree of freedom positioned between the free ends of two identical springs as in figure 2.9 but not connected to the springs, is effectively a non-linear spring system. As the mass moves to the left, compressing the left-hand spring, it becomes disconnected from the right hand spring. Vice versa when it moves to the right. Essentially, neither spring acts in tension and if the gap between the springs is large enough, the mass is moving freely without contact with any spring for a short distance around the middle of its motion.

The motion of the mass may be traced and recorded by attaching a potential divider displacement transducer to the mass. This displacement function may be frequency analysed and the frequency and amplitude of the first harmonic and of the third harmonic determined and then from equations 9.7 and 9.12 the constant δ can be found and from equation 9.13 ω_0 may be found and hence C. Therefore, the mathematical spring force function of equation 9.1 may be drawn and compared with the actual one.

It may also be possible to measure still higher harmonics in the displacement function.

(c) A Harmonic Oscillator with Static Friction Damping

A slider of mass M on a horizontal table attached to a spring of constant C, as in figure 9.9, can be made to have static friction damping if the two contact surfaces are covered with a heavy grit sandpaper. The displacement can be recorded with a potential divider displacement transducer and the parameters of that displacement, marked in figure 9.10, determined. Thus ω_0, the decrement of the amplitude per period, the final offset and the static friction coefficient can be derived from the recording of the free vibration. ω_0, thus determined, may be compared with that predicted from the values of C and M or measured in the absence of damping. Also, the static friction coefficient μ may be derived either from the decrement or from the final offset and compared with the statically determined value (see experiment (a) in section 3.9).

Problems

9.1 Two identical springs of constant 10 kN/m are mounted on a horizontal table along the x-axis and the outer ends are rigidly fixed. There is a gap between the other two ends. A mass of 100 kg is placed in this gap and when it is at the mid-point of the gap it is 1 cm from each of the ends of the springs. The mass is then moved 2 cm to the right and released from rest. Find the equation of the displacement of the mass about the mid-point.

9.2 Approximate the system of problem 9.1 with an identical mass and a non-linear spring which exerts a force of 49.6 N at 1 cm displacement and a force of 50 N at 2 cm displacement. Write down the expression for the effective spring function and find the amplitudes of the fundamental and third harmonic of the free motion.

9.3 What will the steady-state displacement of the mass in the approximate system of problem 9.2 be when it is acted on by the force $200 \cos 50t$ N?

9.4 A mass of 1 kg is attached to two springs of constant 50 N/m in an arrangement like that of figure 2.9. Two rigid stops prevent the mass from travelling beyond 1 cm on either side of the mid-point. Assume that the coefficient of restitution of the stops to be unity, that is, complete reversal of momentum at the stops, and find the equation of the displacement of the mass if the total energy in the system is 0.01 J.

9.5 Approximate the system of problem 9.4 with an identical mass and a non-linear spring which exerts a force of 100 N at 1 cm displacement and 49.6 N at 0.5 cm displacement. Find the equation for the effective force/displacement relationship of the system.

9.6 Draw a graph of the velocity versus the displacement of the mass in problem 9.4.

9.7 The damped free oscillation of a 10 kg mass is recorded, x versus t. When $\ln x$ is plotted versus t the initial slope is -0.8 1/s while the final slope is -0.08 1/s. If the initial displacement is 1 cm and the initial velocity is zero, while the later period is 0.5 s, find the hysteretic damping coefficient and write down the expression for the displacement of the mass.

9.8 Draw a graph of velocity versus displacement for some of the early cycles of the vibration of the mass in problem 9.7.

9.9 If the system of problem 9.7 is subjected to a sinusoidal force cos $6t$ N, what is the steady-state displacement of the mass?

9.10 If the oscillator of problem 9.7 is driven by a force cos ωt N, draw a graph of the average power consumed by the system versus ω and find the resonance frequency, the half-power frequencies and the Q-value of the system.

9.11 A spring of constant 3 N/m is attached to a mass of 5 kg on a horizontal table like that in figure 2.1. The mass is pulled out 10 cm and released from rest. Find the hysteretic damping coefficient. After 50 cycles the vibration is essentially harmonic of period 7.5 s. Write down the expression for the displacement of the mass.

9.12 Find the resonant frequency and the half-power frequencies for the system in problem 9.11.

9.13 A mass/spring system such as that in figure 2.1 is observed to execute a vibration of period 2 s. The amplitude of the movement declines from 5 cm at release from rest to 4 cm after the first period. If the mass is 1 kg, find the spring constant, the static friction coefficient and where the mass finally comes to rest. How many complete oscillations does the mass achieve before coming to rest?

9.14 A horizontal mass/spring oscillator is given an initial displacement and takes 5 s to reach its resting condition with the mass displaced by 0.1 mm. 12 cycles of oscillation occur before the system comes to rest. Find the coefficient of static friction.

9.15 If the mass in the previous problem is increased by a factor of 4, given an initial displacement of 1 cm and released from rest, derive the equations describing the motion of the mass.

9.16 Draw a graph of the velocity versus the displacement of the mass in problem 9.15.

Further Reading

Dimarogonas, A.D., *Vibration Engineering* (West, St. Paul, 1976)
Dinca, F. and Teodosiu, C., *Non-Linear and Random Vibrations* (Academic, New York, 1975)
Seto, W.W., *Theory and Problems of Mechanical Vibrations* (Schaum, New York, 1964)

Steidel, R.F., Jr., *An Introduction to Mechanical Vibrations,* 2nd edition (Wiley, New York, 1979)

Timoshenko, S., Young, D.H. and Weaver, W. Jr., *Vibration Problems in Engineering*, 4th edition (Wiley, New York, 1974)

10 Vibration Measurement and Analysis

10.1 Introduction and Objectives

The aim of this chapter is to describe the electronic techniques used to detect, measure and diagnostically analyse mechanical vibrations.

After reading this chapter the student should be able to:

(a) outline the main sets of reasons why vibrations in mechanical systems are measured;
(b) describe schematically the instruments used to measure vibrations;
(c) describe the transducers used to detect the vibration of a surface;
(d) discuss the frequency response of these transducers;
(e) discuss the analysis of periodic vibrations;
(f) describe the approach to the mathematical analysis of random vibrations including the significance of the autocorrelation function;
(g) outline the approach to analysing shock vibration.

10.2 Effects of Vibrations in Mechanical Systems

There are a number of weighty reasons for the widespread interest in the fundamentals and practical aspects of mechanical vibrations.

One such reason is the possibility of undesirable effects by vibrations on mechanical systems. Any general mechanical system, for example, whole buildings, instruments on the bench in the laboratory, complex mechanical tools on the floor of a workshop, transport vehicles or a human being may be represented by some pattern or form of interconnected mass/spring/damper elements as shown in figure 10.1. Since most driving forces $f(t)$ may be deemed to have harmonic components, the possibility of exciting resonances within the over-all system is great. If a resonance is not damped, the displacement of the mass and hence the stretching of the spring element will tend towards infinity. The spring component will fracture and for this reason undamped resonances must be avoided for the protection of equipment and instruments. This applies also when the human body is part of the over-all system which might experience the damaging resonance. This aspect will be discussed in chapter 12.

VIBRATION MEASUREMENT AND ANALYSIS

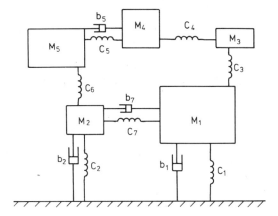

Figure 10.1 In general a mechanical structure may be represented by a set of masses interconnected by springs and energy-absorbing damper elements.

Long-term exposure of a mechanical system to vibrations of frequencies away from resonance can also cause damage through the mechanism of fatigue. Thus, if a mechanical component such as a spring is subjected to repetitive or cyclical applications of stress levels much lower than the ultimate strength, it will fracture after a large number of repetitions of this stress. Indeed, as shown in figure 10.2, if the number of cycles of stress is increased, the amplitude of the stress needed eventually to cause fracture

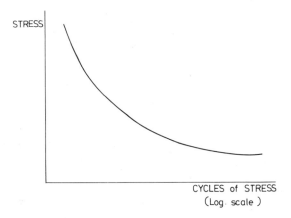

Figure 10.2 Fatigue failure or fracture occurs after a large number of cycles of relatively low stress. The higher the repeated stress the lower the number of cycles before fracture occurs.

becomes lower. The underlying mechanism in fatigue appears to be the gradual unzipping of intermolecular bonds starting from a defect or weakness in the molecular structure.

Another undesirable effect of vibrations is the fact that they can impair the normal functioning of instruments. Thus, if there are vibrations within an electron microscope which magnifies by over $\times 10^4$, a blurred image can result. Vibrations in a microtome can result in cuts of different thicknesses. Likewise, many devices in fine engineering and optics cannot tolerate excessive vibrations. Electrical connections can be undone by vibrations.

Unwanted vibrations in a system, furthermore, indicate inefficiency. Energy is wasted in exciting the vibrations instead of being effectively directed to the work of the system.

Another undesirable side-effect of vibrating structures is the generation of audible noise. Such noise can be psychologically annoying to human beings working in the environment and can render normal voice communication impossible. If extreme, noise can irreparably damage human hearing. The most thorough way of suppressing such noise is to reduce or eliminate the vibrations causing it.

Considerable effort is devoted to the measurement and examination of seismic vibrations associated with earthquakes. These measurements are a vital link in providing advance warning and protection to populations against volcanic eruptions with which are associated earth tremors.

Another area of interest in vibration quantification is the so-called planned or preventive maintenance of equipment, particularly rotating machinery. As this type of machinery ages and undergoes wear, the associated unwanted vibrations tend to become greater. Regular vibration measurement can provide in-service indices of the degeneration of the machinery. Repair or replacement can then be carried out before catastrophic failure and at a time convenient for the factory or plant.

The first step in any of these areas of vibration science is to measure the vibrations in question.

10.3 Measurement Equipment

The most generally used methods of measuring vibrations are electrical. The basic components in such methods are shown schematically in figure 10.3. The key component is the vibration transducer which produces an electrical voltage or current proportional to some quantity in the mechanical vibration, the displacement, velocity or acceleration. Thereafter, a variety of electronic components can carry out any of a range of standard electronic signal processing steps on the vibration voltage. Typical steps include amplification, attenuation, filtering, differentiation and integration. Then the processed signal is measured with a meter,

VIBRATION MEASUREMENT AND ANALYSIS

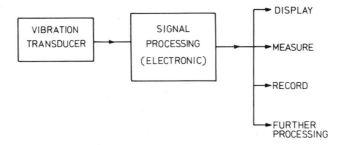

Figuree 10.3 Schematic diagram of a system for detecting and measuring vibrations. It consists of a transducer, an electronic signal processing stage, possibly a recording stage and finally a measurement and analysis stage.

displayed on an oscilloscope, recorded on a chart recorder or tape recorder or further processed and analysed by digital computer.

Attention in this chapter is directed to the nature and characteristics of the transducers and to the main types of analyses used on vibration signals.

10.4 Vibration Transducers, Accelerometer and Vibrometer

A vibration transducer commonly takes the form shown schematically in figure 10.4. It consists of a mass attached to a spring both within a rigid container. The upper end of the spring is attached to the container and the motion of the mass is monitored in some fashion. The whole container is attached to the surface whose vibrations are to be measured.

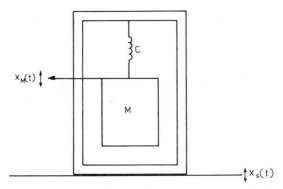

Figure 10.4 A schematic representation of the operating elements of a vibration transducer detecting the displacement of a surface $x_s(t)$.

Assume that the displacement of the surface and hence of the upper end of the spring is pure harmonic thus

$$x_s = A_s \cos \omega t \, . \tag{10.1}$$

This motion of the upper end of the spring results in motion of the mass. Assume this latter motion to have the form

$$x_m = A_m \cos \omega t \, . \tag{10.2}$$

The force equation for the mass M then is

$$M\ddot{x}_m + C(x_m - x_s) = 0 \tag{10.3}$$

or

$$\ddot{x}_m + \omega_0^2(x_m - x_s) = 0 \, . \tag{10.4}$$

Substituting the expressions for x_m and x_s from equations 10.2 and 10.1 respectively, equation 10.4 becomes

$$-\omega^2 A_m \cos \omega t + \omega_0^2(A_m - A_s) \cos \omega t = 0$$

or

$$A_m(\omega_0^2 - \omega^2) = A_s \omega_0^2 \tag{10.5}$$

from which

$$A_m = \left(\frac{1}{1 - (\omega/\omega_0)^2}\right) A_s \, . \tag{10.6}$$

This is the condition for equation 10.2 to describe the motion of the mass. Let

$$r = \frac{\omega}{\omega_0}$$

$$= \frac{f}{f_0} \tag{10.7}$$

the ratio of the frequency of the vibration to the undamped free oscillation frequency of the mass/spring system of the transducer. Then equation 10.6 may be simply written

$$A_m = \left(\frac{1}{1 - r^2}\right) A_s \, . \tag{10.8}$$

VIBRATION MEASUREMENT AND ANALYSIS

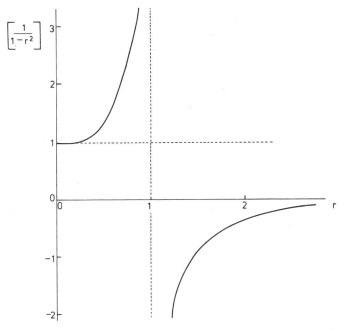

Figure 10.5 The ratio of mass displacement amplitude to surface displacement amplitude as a function of frequency ratio $r\ (= f/f_0)$ for the vibration transducer of figure 10.4.

The function multiplying A_s in this equation is plotted against r in figure 10.5. Without discussing this diagram in detail it is clear that

$$A_m = A_s \text{ for } r \ll 1 \tag{10.9}$$

and the output signal, the displacement of the mass, is then exactly the same as the motion of the surface whose vibration is to be measured. Note

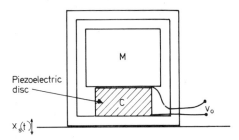

Figure 10.6 Simplified actual construction of a vibration transducer using a piezoelectric sensing element.

that this condition holds when $\omega_0 \gg \omega$ or when the mass M is very small or when the spring constant C is large, that is, the spring is very stiff.

In practice, the spring in the system of figure 10.4 is provided by a disc of piezoelectric material such as barium titanate. Thus, as shown in figure 10.6, the output voltage from across the crystal v_o is proportional to the strain of the piezoelectric crystal spring thus

$$v_o = k(x_m - x_s)$$
$$= k\left(\frac{1}{1-r^2} - 1\right)A_s \cos \omega t$$
$$= k\left(\frac{r^2}{1-r^2}\right)A_s \cos \omega t$$
$$= V_o \cos \omega t \ . \tag{10.10}$$

Here k is a constant. The frequency dependent factor $[r^2/(1 - r^2)]$ is plotted against r in figure 10.7.

Two cases of especial interest can arise.

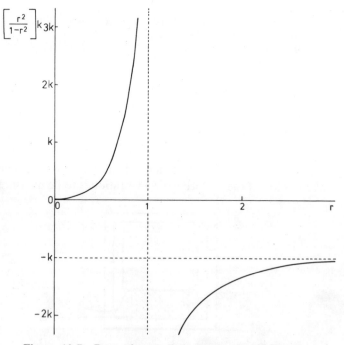

Figure 10.7 Dependence of the sensitivity V_o/A_s of the piezoelectric vibration detector on the frequency ratio $r \ (= f/f_0)$.

VIBRATION MEASUREMENT AND ANALYSIS

(a) r large

If r is large, or $\omega_0 \ll \omega$, the frequency of the vibration to be measured is much greater than the transducer natural frequency. Then

$$\frac{r^2}{1-r^2} = -1$$

and

$$v_o = -kA_s \cos \omega t$$

$$= -kx_s . \qquad (10.11)$$

The output voltage is directly proportional to the displacement of the vibrating surface but is π rad out of phase with it. Such a transducer is called a vibrometer.

(b) r small

If r is small, or $\omega_0 \gg \omega$, the natural frequency of the transducer is much greater than the frequency of the vibration to be measured. Then

$$\frac{r^2}{1-r^2} \rightarrow r^2$$

$$= \left(\frac{\omega}{\omega_0}\right)^2$$

and

$$v_o = k\left(\frac{\omega}{\omega_0}\right)^2 A_s \cos \omega t$$

$$= \left(\frac{k}{\omega_0^2}\right)\omega^2 A_s \cos \omega t$$

$$= -\left(\frac{k}{\omega_0^2}\right)\ddot{x}_s . \qquad (10.12)$$

The output voltage is in this case directly proportional to the acceleration of the vibrating surface but π rad out of phase with it. Such a transducer is called an accelerometer.

The introduction of slight damping into either of these transducers does not greatly affect the performances when the frequency of the vibration is very much less than the natural frequency of the accelerometer, or very much greater than the natural frequency of the vibrometer. If these conditions are relaxed, however, errors can arise due to damping and equations 10.11 and 10.12 do not accurately describe the performances of the transducers.

In selecting an accelerometer for a particular measurement, it is first of all essential that the natural frequency of the transducer be much higher than any frequencies it is called upon to detect but there is a further requirement which arises from that most fundamental law of measurement — that the measuring apparatus affect the system to be measured only to a negligible extent. The accelerometer has a mass M_A that must be rigidly attached to the vibrating mass M_S making the effective mass of the vibrating system $M_S + M_A$. If the measured acceleration is $\ddot{x}_m$, the accelerating force is $(M_S + M_A)\ddot{x}_m$. This can be taken as equal to the accelerating force in the absence of the accelerometer when the acceleration, the true acceleration, is x_t. Then

$$(M_S + M_A)\ddot{x}_m = M_S\ddot{x}_t$$

or

$$\ddot{x}_m = \left(\frac{M_S}{M_S + M_A}\right)\ddot{x}_t . \tag{10.13}$$

If M_A, the mass of the accelerometer, is not much less than the mass of the vibrating structure M_S, the measured acceleration is not the true acceleration $\ddot{x}_t$ and equation 10.13 must be used to adjust the measured values. The addition of an appreciable M_A might also affect any resonances in the system under test and thereby also change the system unduly. Therefore, for minimal artifacts, an accelerometer of mass much less than the mass of the structure under examination should be chosen.

Eample 10.1

An accelerometer of mass 0.1 kg and resonant frequency 100 Hz registers the following voltage when attached to a mass of 10 kg: $v_o = 0.01 \cos 500t$ V. What is the true acceleration amplitude? What is the r.m.s. value of the displacement? Take the k constant to be 1.

Solution

Here $\omega_0 = 628$ rad/s and $r = 500/628 = 0.8$. If A_t is the true amplitude of the mass, then from equations 10.13 and 10.10

$$A_t = (10.1/10.)\,[(1 - r^2)/r^2] \times 0.01$$

$$= 1.01 \times 0.56 \times 0.01 = 0.0057 \text{ m}$$

VIBRATION MEASUREMENT AND ANALYSIS

Then

$$A_{\text{r.m.s.}} = 0.004 \text{ m}$$

true amplitude of the acceleration $= \omega^2 A_t$

$$= 1.43 \text{ km/s}^2 .$$

10.5 Other Displacement, Strain and Force Transducers

A simple form of displacement transducer is the sliding wire potential divider. For linear displacements, a straight wire of uniform cross-sectional area, length L and resistance R may be used as shown in figure 10.8. A

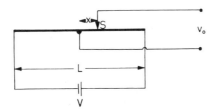

Figure 10.8 A linear potential divider acting as a centre zero displacement transducer. The output voltage v_0, is proportional to x.

D.C. potential drop V is established along the wire. The sliding contact S is rigidly attached to the mass or surface whose displacement is to be monitored. Then the output voltage of the transducer, between the slider and the fixed mid-point of the slide wire, is given by

$$v_o = \frac{V}{L} x . \qquad (10.14)$$

This output voltage is directly proportional to the displacement x, being positive when x is positive and negative when x is negative.

A circular slide wire may be constructed to transduce angular displacement. In both cases it is essential that good contact between slider and slide wire be maintained at all times during the motion. A possible drawback of this transducer is the friction between the contacts during sliding. This factor may appreciably affect the motion being studied.

A very widely used strain transducer is the electrical resistance strain gauge. Basically this is a length L of uniform resistance wire which is bonded to the material undergoing the strain to be measured. Unstrained, the resistance R of the length of wire is

$$R = \frac{\rho L}{A} \tag{10.15}$$

where ρ is the resistivity of the material of the wire and A is the uniform cross-sectional area of the wire. When stretched or shortened with the specimen being strained the length changes to $L + \Delta L$ and the resistance to $R + \Delta R$. A sensitive Wheatstone bridge, operated in a slightly out of balance mode, is used to generate a voltage proportional to fractional change of resistance, thus

$$\begin{aligned} v_o &= k\frac{\Delta R}{R} \\ &= k\frac{\Delta L}{L} \\ &= kS \end{aligned} \tag{10.16}$$

where k is a constant, the sensitivity, and S is the strain. To maximise the fractional resistance change, a conventional construction of a strain gauge is that of figure 10.9 in which a considerable length of resistance wire is fitted into a short space. The plastic backing may be adhesively attached to the surface of the specimen under test.

Figure 10.9 A ten-strand wire strain gauge mounted on a plastic backing which may in turn be adhesively attached to the surface under test. The measured strain is in the left to right direction as shown.

A strain gauge bonded to the surface of a cylindrical specimen of a material of known cross-sectional area A and Young's modulus E may be used to detect the force applied along the axis of this specimen. In this configuration the output voltage may be developed, thus, from equations 1.22 and 1.20

$$\begin{aligned} v_o &= kS \\ &= k\frac{T}{E} \\ &= k\frac{1}{AE}F \\ &= k_1 F \end{aligned} \tag{10.17}$$

VIBRATION MEASUREMENT AND ANALYSIS

where F is the force to be measured and k_1 is a constant. If F is time-varying then v_o will likewise be time-varying.

Strain gauges made of semiconductor materials are probably the most commonly available and, while they offer advantages such as greater sensitivity and miniaturisation, their basic performance equation is that of 10.16.

Example 10.2

A strain gauge mounted longitudinally on a steel rod of radius 2 cm and length 5 cm produces a fractional change in resistance varying in time thus $\Delta R/R = 0.01 \cos 15t$. What is the force being transmitted by the rod?

Solution

The strain in the rod may be written $S = 0.01 \cos 15t$. Therefore the force on the rod, $F = AES$, where A is the cross-sectional area and E is Young's modulus for steel (0.2 GN/m^2). So $F = 2.51 \cos 15t$ MN.

10.6 Analysis of Periodic Vibrations

If the vibration signal from the accelerometer has a single frequency (a pure tone), it is completely specified by the frequency and amplitude, thus

$$v_o = V_o \cos \omega t$$

$$= k\left(\frac{\omega^2}{\omega_0^2}\right) A_s \cos \omega t \ . \tag{10.12}$$

The frequency of the vibration f is given by

$$f = \frac{\omega}{2\pi}$$

and the amplitude of the displacement A_s is found from the amplitude of the measured voltage thus

$$A_s = \frac{1}{k}\left(\frac{\omega_0^2}{\omega^2}\right) V_o \ . \tag{10.18}$$

The amplitude of the acceleration is given by

$$\omega^2 A_s = \frac{\omega_0^2}{k} V_o$$

$$= \frac{1}{k_1} V_o \ . \tag{10.19}$$

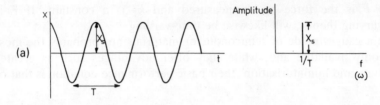

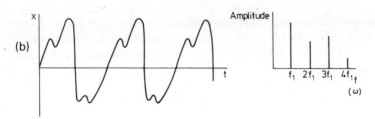

Figure 10.10 (a) A pure tone in the time domain may be represented by a single line of amplitude equal to the amplitude of the sine wave, in the frequency domain, located at $f = 1/T$. (b) A general periodic function in the time domain is represented by a line spectrum in the frequency domain. Each line is located at the frequency of that harmonic and has an amplitude equal to the amplitude of that harmonic in the Fourier series expansion.

Here k_1 is a constant for the accelerometer, the sensitivity of the transducer.

A more common way of describing the magnitude of a pure tone displacement is to use the root mean square value of the displacement, $A_{r.m.s.}$ where

$$A_{r.m.s.} = \sqrt{\left(\frac{1}{T}\int_0^T x_s^2 \, dt\right)}$$

$$= \frac{1}{\sqrt{2}} A_s . \tag{10.20}$$

Figure 10.10 shows such a pure tone plotted on the time axis and on a frequency axis where the line spectrum has but a single line at the pure tone frequency of amplitude A_s.

If the vibratory signal v_o is more generally periodic, it may be Fourier-analysed into its spectral components. The time domain and frequency domain representations for such a signal are shown in figure 10.10(b). An electronic frequency analyser, wave analyser or spectrum

VIBRATION MEASUREMENT AND ANALYSIS

analyser can perform this Fourier analysis on the time domain signal to yield the spectrum. Equations 10.18 and 10.19 may then be used to calculate the spectrum of the actual vibration from that of the electrical signal.

Knowledge of the spectral composition of a vibration can often be used to pinpoint the root causes of different spectral components. The various contributory causes of the complex periodic vibration may then be dealt with, each in the most appropriate way.

10.7 Random Vibrations

In many situations, the sources of the vibrations are not harmonically linked and the summation of their effects on a particular system or instrument might produce a displacement/time relationship as shown in figure 10.11. This is a random vibration. The first step in analysing it is to record a long epoch of the displacement on a paper chart recorder.

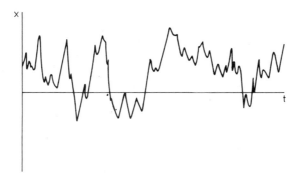

Figure 10.11 A non-periodic, random function of t.

A widespread and tractable sub-group of the random vibrations are the stationary random vibrations whose statistical characteristics are invariant as time goes by. For such vibrations a most useful parameter is the probability density $p(x)$ which may be explained with reference to figure 10.12. This figure is a short record of the random displacement and is of duration $T(s)$. The probability density is defined thus

$$p(x) = \lim_{\Delta x \to 0} \left[\frac{P(x+\Delta x) - P(x)}{\Delta x} \right]$$

$$= \frac{\mathrm{d}P(x)}{\mathrm{d}x} \tag{10.21}$$

where $P(x)$ is the distribution function of the random variable x. $P(x)$ is defined as the probability of x being less than or equal to a given value of x. The probability of having a value of x within the interval Δx from x to $x + \Delta x$ is

$$P(x + \Delta x) - P(x) = \int_x^{x+\Delta x} p(x)\,dx$$

$$= \frac{\Delta t_1 + \Delta t_2 + \Delta t_3 + \Delta t_4}{T}$$

$$= \frac{\Sigma_k \Delta t_k}{T} \qquad (10.22)$$

from figure 10.12. This latter calculation may be readily carried out with the appropriate readings from a chart record of the vibration. Simply choose a measurement epoch T s long and follow the procedure of figure 10.12 and equation 10.22.

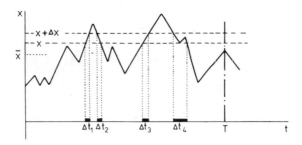

Figure 10.12 Demonstration of how to calculate graphically the probability density function $p(x)$ of the random variable x following equations 10.17 and 10.18.

The mean value of x, $\bar{x}$ may be calculated from the probability density function thus

$$\bar{x} = \int_{-\infty}^{\infty} x\,p(x)\,dx \ . \qquad (10.23)$$

The variance σ^2 of the random variable x may also be derived from the probability density function, thus

$$\sigma^2 = \int_{-\infty}^{\infty} (x - \bar{x})^2\,p(x)\,dx \ . \qquad (10.24)$$

σ, the square root of the variance, is called the standard deviation.

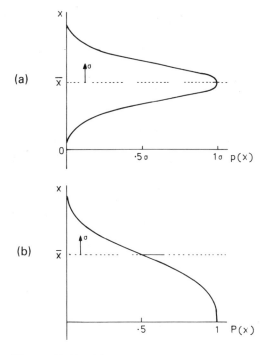

Figure 10.13 (a) A normal or gaussian probability density function of the random variable x of mean value $\bar{x}$ and standard deviation σ. (b) The distribution function $P(x)$ of the random variable x which has a normal probability density function.

Various probability density functions can occur but, for a random vibration, a normal or gaussian distribution applies. In this case

$$p(x) = \left(\frac{1}{\sqrt{(2\pi)}\sigma}\right) e^{-(x-\bar{x})^2/2\sigma^2} \qquad (10.25)$$

a function shown in figure 10.13(a). 68 per cent of all values of x lie within a band $\bar{x} \pm \sigma$, 95 per cent are within the band $\bar{x} \pm 2\sigma$ while 98 per cent are within the band $\bar{x} \pm 3\sigma$.

For such a probability density function the distribution function is

$$P(x) = 0.5 + \mathrm{erf}\left(\frac{x-\bar{x}}{\sigma}\right) \qquad (10.26)$$

which is illustrated in figure 10.13(b).

Another important conclusion can be drawn from a further examination of equation 10.24 in conjunction with equation 10.22.

$$\sigma^2 = \int_{-\infty}^{\infty} (x - \bar{x})^2 \, p(x) \, dx$$

$$= \lim_{\Delta t \to 0} \sum_{0}^{T} \frac{(x - \bar{x})^2 \Delta t}{T}$$

$$= \int_{0}^{T} \frac{(x - \bar{x})^2 \, dt}{T}$$

$$= \frac{1}{T} \int_{0}^{T} (x - \bar{x})^2 \, dt \qquad (10.27)$$

which is the mean square of the A.C. portion of the stationary random vibration. Thus σ, the standard deviation of the normal distribution of x, is numerically equal to the root-mean-square (r.m.s.) value of the A.C. component of x. The r.m.s. value of the electrical vibration signal may be directly measured with a conventional electrical A.C. voltmeter.

The statistical description of the stationary random vibration $x(t)$ disregards all the time or frequency information in the signal. If there is some time-related order embedded in the random fluctuations and due, for example, to the switch-on or switch-off of one of many vibration sources, the statistical approach is incapable of distinguishing it but the extent of such order can be gauged by examination of the autocorrelation function defined thus

$$R(\tau) = \lim_{T \to \infty} \frac{1}{T} \int_{0}^{T} x(t) \, x(t + \tau) \, dt \, . \qquad (10.28)$$

If there happens to be a non-zero mean value of x, $\bar{x}$, a D.C. level, it emerges that

$$R(\tau) = \lim_{T \to \infty} \frac{1}{T} \int_{0}^{T} [x(t) - \bar{x}][x(t + \tau) - \bar{x}] \, dt - \bar{x}^2 \qquad (10.29)$$

where $\bar{x}^2$ is a constant. If there are only A.C. components in $x(t)$ then $\bar{x} = 0$. This case is sufficiently general to need further examination.

The autocorrelation function $R(\tau)$ is the average value of the product of $x(t)$ and $x(t + \tau)$ which is the displacement function shifted in time by τ as shown in figure 10.14. In calculating the autocorrelation function, therefore, the $x(t)$ function serves as a temporal filter which is moved along the time axis past itself. When the superposition between $x(t)$ and $x(t + \tau)$ is good, the value of $R(\tau)$ is high and vice versa when the superposition is a

VIBRATION MEASUREMENT AND ANALYSIS

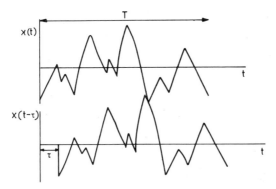

Figure 10.14 The calculation of the autocorrelation function $R(\tau)$ of the random variable $x(t)$ involves shifting the function by τ, multiplying the shifted function by the unshifted original function and calculating the area under the product function. If the original and shifted functions fit together well, that is correlate well, the area will be large. If they do not correlate, the area will be near zero.

poor fit. Thus, if there is no relative shift ($\tau = 0$) there is perfect superposition and $R(\tau)$ is a maximum. Thus

$$R(0) \text{ is maximum.} \tag{10.30}$$

Furthermore from equations 10.28 and 10.27

$$R(0) = \lim_{T \to \infty} \frac{1}{T} \int_0^T x^2(t)\, dt$$
$$= \sigma^2 \tag{10.31}$$

which is the mean square value of the A.C. portion of the $x(t)$ vibration signal and is proportional to the average power in that signal.

In writing the autocorrelation function $R(\tau)$ of equation 10.28, shifting one $x(t)$ by a positive τ is equivalent to shifting the other by $-\tau$ and therefore $R(\tau)$ is an even function

$$R(-\tau) = R(\tau). \tag{10.32}$$

If $x(t)$ is a random vibration signal, then

$$R(\tau) \xrightarrow[T \to \infty]{} 0. \tag{10.33}$$

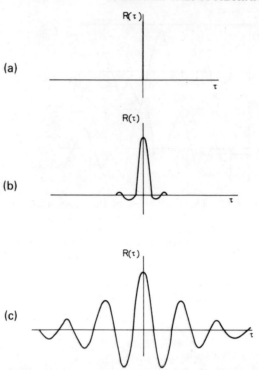

Figure 10.15 Typical autocorrelation functions (a) for a wide-band random variable (b) for an intermediate band-width random variable and (c) for a narrow band-width random variable.

But if there is a D.C. term in $x(t)$

$$R(\tau) \xrightarrow[\tau \to \infty]{} \bar{x}^2. \tag{10.34}$$

If there is a periodicity of T in $x(t)$ then there will be the same periodicity in $R(\tau)$ so that

$$R(\tau \pm T) = R(\tau) \tag{10.35}$$

and it will not tend to zero as τ increases. These latter two features, the possible presence of D.C. or a periodicity, expressed in equations 10.34 and 10.35, allow departures from perfect randomness to be quantified. Such observations can be very useful in diagnosing the causes of vibrations.

Typical autocorrelation functions for different random vibrations are shown in figure 10.15.

VIBRATION MEASUREMENT AND ANALYSIS

The Fourier transform of the autocorrelation function $R(\tau)$, is called the spectral density function $S(\omega)$ of the stationary random vibration $x(t)$. From equation 5.40

$$S(\omega) = \int_{-\infty}^{\infty} R(\tau) e^{-j\omega\tau} d\tau \ . \qquad (10.36)$$

If $S(\omega)$ is known, $R(\tau)$ may be derived from it via the inversion integral of equation 5.41 thus

$$R(\tau) = \frac{1}{2\pi} \int_{-\infty}^{\infty} S(\omega) e^{j\omega\tau} d\omega \ . \qquad (10.37)$$

Since for physical vibrations $\omega \geq 0$ only and, from equation 10.32, $R(\tau) = R(-\tau)$, equation 10.37 becomes

$$R(\tau) = \frac{1}{\pi} \int_{0}^{\infty} S(\omega) \cos \omega\tau \, d\omega$$

$$= \int_{0}^{\infty} Q(2\pi f) \cos(2\pi f\tau) \, df \qquad (10.38)$$

where

$$Q(2\pi f) = 2S(2\pi f) \ . \qquad (10.39)$$

At $\tau = 0$, equation 10.38 yields

$$R(0) = \int_{0}^{\infty} Q(2\pi f) \, df$$

$$= \lim_{T \to \infty} \frac{1}{T} \int_{0}^{T} x^2(t) \, dt$$

$$= \sigma^2 \qquad (10.40)$$

from equation 10.31. Both of these integrals are measures of the average power in the random vibration, in the one case measured in the time domain and, in the other, measured in the frequency domain.

The spectral density function $S(2\pi f)$ or $\frac{1}{2}Q(2\pi f)$ can be measured with a frequency analyser as indicated in figure 10.16. If the bandwidth of the filter in the analyser Δf is small then

$$\int_{f}^{f+\Delta f} Q(2\pi f) \, df = Q(2\pi f) \, \Delta f \ . \qquad (10.41)$$

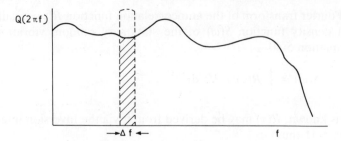

Figure 10.16 The spectral density function $S(\omega)$ or $S(2\pi f)$, as measured with a frequency analyser with filter bandwidth Δf, can be multiplied by 2 to yield $Q(2\pi f)$. In turn, following equation 10.38, this function may be used to calculate the variance or the standard deviation of the probability density function of the random variable.

In the limit of $\Delta f \to 0$, from equation 10.40

$$Q(2\pi f) = \lim_{\Delta f \to 0} \lim_{T \to \infty} \frac{1}{T\Delta f} \int_0^T x^2(t)\, dt$$

$$= \lim_{\Delta f \to 0} \frac{\sigma^2}{\Delta f}. \qquad (10.42)$$

Therefore the area under the graph of figure 10.16 gives the variance of the probability density function for the random vibration $x(t)$ This is also the average power of the vibration or the mean square value of $x(t)$.

Thus the mean square value of the stationary random vibration $x(t)$ is a measure of the average power and of the statistics of the vibration. In fact a stationary random process can be defined as a process for which the variance σ^2 or the standard deviation σ is constant.

One final aspect of stationary random vibrations is noteworthy. If the system is linear, and if the input driving force is a stationary random process, then so also is the output vibration.

Example 10.3

What is the probability density function for the variable $x = [-1 + (2/T)t]U(t - nT)$, $n = 0, 1, 2 \ldots$ the periodic saw-tooth function?

VIBRATION MEASUREMENT AND ANALYSIS

Solution

From equations 10.21 and 10.22 the probability density function $p(x)$ may be found as follows

$$dP(x) = \frac{1}{T}\Sigma \Delta t$$

$$= \frac{1}{T} dt$$

$$= \frac{1}{T} \frac{T}{2} dx .$$

But

$$p(x) = \frac{dP(x)}{dx}$$

$$= \tfrac{1}{2} \text{ for } -1 \leq x \leq 1 .$$

Example 10.4

What is the autocorrelation function of the variable x of example 10.3?

Solution

$$R(\tau) = \frac{1}{T}\int_0^T \left(-1 + \frac{2}{T}t\right)\left(-1 + \frac{2}{T}(t + \tau)\right) dt$$

$$= \frac{1}{T}\int_0^T \left[1 - \frac{2}{T}(2t + \tau) + \frac{4}{T^2}(\tau t + t^2)\right] dt$$

$$= \frac{1}{3} .$$

Since this is constant it is the value of $R(0)$ and also the value of σ^2 as may be derived from

$$\sigma^2 = \int_{-1}^{1} x^2 \, p(x) \, dx$$

$$= \tfrac{1}{2}\int_{-1}^{1} x^2 \, dx$$

$$= \frac{1}{3} .$$

10.8 Shocks/Pulsed Vibrations

As discussed in chapter 5, an isolated shock or pulsed vibratory signal may be analysed harmonically. A rectangular pulse has a continuous spectrum as shown in figure 5.7. Pulses of other shapes in the time domain have differently shaped continuous spectra. In general, the sharper the pulse the more appreciable the high frequency components in the spectrum and the wider the bandwidth of the pulse spectrum.

In measuring such a pulsed vibration it is therefore difficult to achieve in the accelerometer the basic requirement that the natural frequency of the transducer be much greater than the maximum frequency to be measured; $\omega_0 \gg \omega$. A very high natural frequency accelerometer must be chosen but some degradation of the high frequency part of the spectrum is inevitable. Such degradation can be reduced by incorporating some damping into the accelerometer so that the resonance peak of the transducer response, shown in figure 10.7, is kept as low as possible.

In spite of the inaccuracy in the high frequency tail of the spectrum, due to the limited accelerometer frequency response, spectral analysis of the lower range of the shock spectrum does allow broad details of the shock to be described by comparison with the spectra of known types of shock, such as that of figure 5.7. This can permit some extrapolation to high frequencies.

10.9 Experiments

(a) Frequency Response of Vibration Transducers

An electrically driven vibrator of calibrated output force amplitude/frequency characteristic can serve as the vibration source. The accelerometer is attached rigidly to the vibrating surface and the output voltage of the accelerometer monitored on an oscilloscope. At each frequency this output voltage is divided by the amplitude of the acceleration of the surface to give the sensitivity of the transducer. The graph of sensitivity versus frequency is the frequency response of the accelerometer. A similar experiment may be done with a vibrometer except that in this case the output voltage divided by the surface displacement amplitude is the sensitivity.

(b) Frequency Response of the Electronics of Vibration Measurement Apparatus

A signal generator may be used to supply a voltage input to the electronics, instead of the transducer, and the output voltage monitored on an

VIBRATION MEASUREMENT AND ANALYSIS

oscilloscope. The ratio of output voltage amplitude to input voltage amplitude is the response of the system. This response is plotted versus frequency to give a complete picture of the performance of the electronics.

(c) Measurement and Analysis of Vibrations

A typical vibration detective problem is to assign responsibility for the various frequency components in the vibration of a specific surface to different vibration sources in the environment, the same room or other parts of the building, etc. The first step is to attach the accelerometer rigidly to the surface in question and to measure and frequency-analyse the vibration there. Then, if possible, each prime mover machine in the vicinity is switched off in turn and the spectrum of the vibration again measured. The changes in the spectrum at each step allow one to deduce the sources and their importance in the over-all vibration. Many different locations in a workshop or laboratory may be investigated in this way.

If a major portion of the vibrations originates elsewhere in the building, as in a basement boiler-room or a machine-shop, it may be necessary to bring the vibration measurement and analysis equipment to such sources to characterise the spectra of the vibrations produced by them. From such spectra, and from an inspection of the vibration transmission pathways from the sources to the target region of interest, it may be possible to assign responsibility for the observed vibrations.

(d) Measurement of Shocks

Providing the frequency response of the accelerometer and the electronics is wide enough, essentially continuous spectrum vibrations, such as repeated hammer-blows, can be detected and frequency-analysed. Indeed, if a spectrum analyser is available, one-shot pulses can be analysed directly. An interesting application can arise for such shock measurement and analysis if drop-forging or pile-driving is going on in the vicinity.

Problems

10.1 A potential divider displacement transducer has a mass of 0.01 kg and a damping coefficient of 2 kg/s. When attached to a vibrating mass, which is associated with a spring of constant 22.7 N/m, the period measured is 0.54 s. What are the true mass and the true period?

10.2 An accelerometer of total mass 0.02 kg, of resonant frequency 600 Hz and of transduction factor (k) 5×10^6 is used to measure the vibration of a mass of 0.5 kg. The output voltage $v_o = 5 \cos 45t + \cos 90t$ V. What are the true acceleration and displacement of the mass?

10.3 An accelerometer has a mass of 0.1 kg and a spring of constant 0.1 MN/m. What maximum frequency vibration will the accelerometer measure to an accuracy of 1 per cent and to an accuracy of 5 per cent?

10.4 A vibrometer has a natural frequency of 2 Hz. In measuring a vibration of frequency 10 Hz, the measured amplitude is 0.5 cm. What is the true amplitude?

10.5 What is the minimum frequency of vibration that can be measured to an accuracy of 2 per cent with the vibrometer of problem 10.4?

10.6 In relation to each of the following functions, calculate the amplitude, the r.m.s. value, the period and the phase angle.

(a) $x = 10 \cos 4t - 5 \sin 4t$;
(b) $x = 0.01 \sin t + 0.05 \cos t$;
(c) $x = -50 \cos 50t + 50 \sin 50t$;
(d) $x = 7 \cos 200t - 9 \sin 200t$.

10.7 Calculate the probability density function for the variable $x = |\sin 2t|$.

10.8 What are the mean and variance of the variable in problem 10.7?

10.9 Calculate the autocorrelation function for the variable

$$x = |\sin 2\pi t| \qquad 0 \leq t \leq 0.5\pi$$
$$= 0 \qquad 0.5\pi \leq t \leq 1$$

which is repeated with period π s. What is the power spectrum of this variable?

10.10 Calculate the autocorrelation function for the rectangular pulse train. Each pulse is of amplitude 1, of duration τ s and the period is T s.

10.11 Repeat problem 10.10 for a square wave of period T and amplitude unity.

10.12 If the spectral density of a vibration can be graphically described by an isosceles triangle of height A and base $2f_m$, centred on the zero of the frequency axis, find the autocorrelation function of the variable as well as the variance.

10.13 If the spectral density of a vibration can be represented $S(f) = 10$, for $|f| \leq 100$ Hz and otherwise zero, find the probability density function of the vibration displacement x.

10.14 White noise random vibration has a spectral density function constant at 100 at all frequencies from $-\infty$ to $+\infty$. What is the autocorrelation function for this white noise?

Further Reading

Beranek, L.L., *Noise and Vibration Control* (McGraw-Hill, New York, 1971)
Bramer, T.P., et al., *Basic Vibration Control* (Sound Research Laboratories, Sudbury, 1977)
Broch, J.T., *Mechanical Vibration and Shock Measurement* (Brüel & Kjaer, Naerum, 1973)
Clayton, G.B., *Operational Amplifiers* (Butterworths, London, 1974)
Cook, N.H., and Rabinowitz, E., *Physical Measurement and Analysis* (Addison-Wesley, Reading, MA., 1963)
Cooper, W.D., *Electronic Instrumentation and Measurement Techniques* (Prentice-Hall, Englewood Cliffs, NJ., 1970)
Doebelin, E.O., *Measurement Systems: Application and Design*, 2nd edition (McGraw-Hill, New York, 1975)
Dinca, F., and Teodosiu, C., *Nonlinear and Random Vibrations* (Academic, New York, 1975)
Graeme, J.G., *Applications of Operational Amplifiers* (McGraw-Hill, New York, 1974)
Harvey, G.F. (Ed.), *Transducer Compendium*, 2nd edition (Plenum, New York, 1969)
Harris, C.M. (Ed.), *Handbook of Noise Control* (McGraw-Hill, New York, 1957)
Herceg, E.E., *Handbook of Measurement and Control* (Schaevitz Engineering, Pennsauken, NJ., 1972) .
Hobbie, R.K., *Intermediate Physics for Medicine and Biology* (Wiley, New York, 1978)
Lion, K.S., *Elements of Electrical and Electronic Instrumentation* (McGraw-Hill, New York, 1975)
Neubert, H.K.P., *Strain Gauges: Kinds and Uses* (Macmillan, London, 1967)
Norton, H.N., *Handbook of Transducers for Electronic Measuring Systems* (Prentice-Hall, Englewood Cliffs, NJ., 1969)
Papoulis, A., *Signal Analysis* (McGraw-Hill, New York, 1977)
Petrusewicz, S.A., and Longmore, D.K., (Eds.), *Noise and Vibration Control for Industrialists* (Elek Science, London, 1974)
Stein, P.K., *Measurement Engineering, Vol. 1: Basic Principles* (Engineering Services, Tempe, AZ., 1962)
Stout, D.F., and Kaufman, M., *Handbook of Operational Amplifier Circuit Design* (McGraw-Hill, New York, 1976)
Taub, H., and Schilling, D.L., *Principles of Communication Systems* (McGraw-Hill, New York, 1971)
Wart, J.V., Huelsman, L.P., and Korn, G.A., *Introduction to Operational Amplifier Theory and Applications* (McGraw-Hill, New York, 1975)

11 Vibration Isolation and Control

11.1 Introduction and Objectives

The aim in this chapter is to describe the ways and means of reducing vibrations in prime movers, isolating sources of vibration from other vulnerable structures and protecting devices from vibrations in the environment.

After reading this chapter the student should be able to:

(a) show how eccentricities in prime movers can produce vibrations;
(b) describe how balancing can reduce vibrations at source;
(c) outline how to isolate a source of vibrations to prevent the transmission of these vibrations to other structures;
(d) design passive mountings to protect equipment from vibrations;
(e) design dynamic vibration absorbers to protect structures from vibrations.

11.2 Sources of Vibrations

Resonant vibrations can be stimulated in mechanical systems by shocks or impact forces such as hammer blows. Frictional forces between a prime mover and the system of interest can also induce resonance but one of the most common sources of harmonic vibrational forces is an unbalanced rotor in a rotating machine.

Take as an example a cylindrical rotor of mass M rotating about its long axis at ω rad/s. The axis of rotation through C in figure 11.1 is eccentric to the true axis through O by e m. The system is equivalent to a mass M concentrated along the O axis rotating at ω rad/s about the C axis. This exerts a centrifugal force on the end bearings

$$F_c = Me\omega^2 \ . \tag{11.1}$$

This force acts radially relative to C as shown in figure 11.1. Its vertical component $F_{c(V)}$, which is the vertical force on the bearings and in turn the vertical force on the bearing mounts, is

VIBRATION ISOLATION AND CONTROL

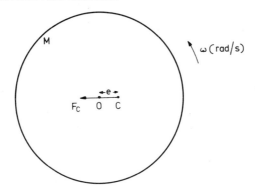

Figure 11.1 A cylindrical rotor rotating about an axis through C but with centre of mass along the axis through O, a distance e, the eccentricity, from C. The rotation is counter-clockwise at ω rad/s.

$$F_{c(V)} = Me\omega^2 \cos \omega t . \tag{11.2}$$

This is a pure harmonic force of frequency $f = \omega/2\pi$ and amplitude $Me\omega^2$. The horizontal component of this force $F_{c(H)}$ is also a pure sinusoid

$$F_{c(H)} = Me\omega^2 \sin \omega t. \tag{11.3}$$

Since some degree of eccentricity is virtually impossible to avoid in rotational machinery, motors, generators, pulleys, flywheels, etc., such prime mover machinery is always accompanied by some pure tone forces.

A number of features of these forces should be borne in mind. The frequency is directly proportional to the speed of rotation. The amplitude of the force is directly proportional to the mass of the rotor, to the square of the rotational speed and to the eccentricity e of the mounting.

Rotation of a long flexible rotor as in figure 11.2 can also result in sinusoidal forces on the bearings. When the rotor bends slightly under its own weight the centre of mass becomes eccentric or off-axis. The greater the bend, the greater the eccentricity. In this case, when the rotor rotates, vertical and horizontal forces act on the bearings formally similar to those described in equations 11.2 and 11.3. The centrifugal force at the mid-point of the length of the rotor also increases the eccentricity as predicted by equation 1.25 describing the bending of a beam. Thus, the greater the rotational speed, the greater the eccentricity and so the amplitudes of the horizontal and vertical vibratory forces on the bearings rise more sharply with rotational speed than the square function of equations 11.2 and 11.3.

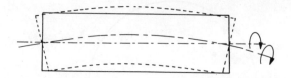

Figure 11.2 A long flexible rotor which sags during rotation has its centre of mass effectively rotated eccentrically to the axis of rotation.

Example 11.1

A flywheel of mass 1000 kg rotating at 500 revolutions per minute is found to exert a maximum horizontal force of 600 N on the bearings. What is the eccentricity of the mounting?

Solution

$$\omega = 2\pi f = \frac{2\pi}{T} = \frac{2 \times 500}{60} = 52.4 \text{ rad/s} .$$

From equation 11.1

$$F_{c(H)}(\max) = M\omega^2 e = 600 \text{ N, so}$$

$$e = \frac{600}{1000 \times (52.4)^2} = 0.22 \text{ mm} .$$

11.3 Vibration Prevention

The harmonic force of equation 11.2 would be eliminated if the eccentricity e were zero. To achieve this criterion, very fine precision-machining of all parts of the rotor and bearings is needed. The more precise the machining, the more expensive the equipment. Then after some wear in service, the eccentricity would be expected to increase anyhow. Therefore the procedure of balancing is often applied to minimise the force amplitude.

The basic idea in balancing the rotor is to mount a second small mass M_B on the rotor but across the diameter from O through C at B on the rim of the rotor in figure 11.3 such that

$$M_B r = Me \tag{11.4}$$

VIBRATION ISOLATION AND CONTROL

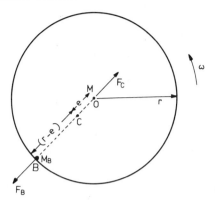

Figure 11.3 If a balancing mass M_B is attached to the rim of the eccentrically rotating rotor of figure 11.1 along the same diameter as the centre of mass O and the centre of rotation C, but on the opposite side from the centre of mass, it produces a centrifugal force F_B opposite to the eccentric rotational force F_C.

where r is the radius of the rotor. Then when the rotor rotates both of these eccentric masses, the rotor mass itself M and the balancing mass M_B exert oppositely directed centrifugal forces so that the total or net centrifugal force is

$$F_c = Me\omega^2 - M_B r\omega^2$$
$$= 0 \ . \tag{11.5}$$

Therefore the net horizontal force and the net vertical force are both zero.

In practice the achievement of balance is a more detailed procedure than the static balancing shown in figure 11.4 because the masses, especially that of the rotor, are not concentrated at points and the rotor may have flexibility along its length and hence a varying effective eccentricity along the length. Dynamic balancing with a number of balance masses positioned at a variety of radial and angular positions must often be used. The positions of the masses are fine-tuned between rotation tests at the working rotational speed until minimum unbalanced rotor force is achieved.

In most practical situations it is not possible to completely eliminate the vibratory force at source. Measures may then be taken to minimise the transfer of such forces to neighbouring systems. The vibratory force source may be isolated.

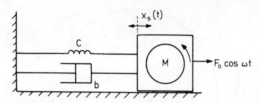

Figure 11.4 A machine of mass M on a spring plus damper mounting. The machine has an internally generated harmonic force which causes the mass to vibrate.

Example 11.2

What balancing mass located 1 m from the axis of rotation would balance the eccentric mounting of the flywheel of example 11.1? How far from the axis should a 1 kg mass be placed for balance?

Solution

(a) For balance, $M_B \times 1 = Me = 1000 \times 0.22 \times 10^{-3}$ and $M_B = 0.22$ kg located 1 m on the opposite side of the diameter through the axis and the true central axis.

(b) At balance, $1 \times r_b = Me$, so $r_b = 0.22$ m.

11.4 Isolation of Vibration Source

Consider the situation depicted in figure 11.4 where a body of mass M experiences a harmonic driving force $F_0 \cos \omega t$ due, for instance, to an unbalanced rotor within the body. In turn, the body is attached to the vertical wall by means of a mounting characterised by a spring constant C and a damping coefficient b.

The issue to be addressed here is how effectively is the vibratory force being applied to the mass in turn transmitted to the vertical wall. The force on the wall in the steady-state is the sum of the spring force and the damper force

$$f_w = Cx_s + b\dot{x}_s \tag{11.6}$$

where x_s is the steady-state displacement of the mass in the horizontal direction. x_s is given by equation 4.8 .

VIBRATION ISOLATION AND CONTROL

$$x_s = C_s \cos(\omega t - \gamma_s). \qquad (4.8)$$

The velocity is

$$\dot{x}_s = -\omega C_s \sin(\omega t - \gamma_s) \qquad (11.7)$$

and the amplitude C_s is given by equation 4.18

$$C_s = \left(\frac{(F_0/M)}{\sqrt{[(\omega_0^2 - \omega^2)^2 + (2\alpha\omega)^2]}} \right) \qquad (11.8)$$

while the phase angle γ_s is given by equation 4.19

$$\gamma_s = \tan^{-1}\left(\frac{2\alpha\omega}{\omega_0^2 - \omega^2} \right). \qquad (4.19)$$

Substituting the expression for the displacement x_s and the velocity $\dot{x}_s$ of equations 4.8 and 11.7 respectively into equation 11.6 produces

$$f_w = C_s[C \cos(\omega t - \gamma_s) - b\omega \sin(\omega t - \gamma_s)]$$

$$= C_s \sqrt{(C^2 + b^2\omega^2)} \cos(\omega t - \gamma_s + \varphi) \qquad (11.9)$$

where

$$\varphi = \tan^{-1}\left(\frac{b\omega}{C} \right)$$

$$= \tan^{-1}\left(\frac{2\alpha\omega}{\omega_0^2} \right). \qquad (11.10)$$

Substituting from equation 11.8 the expression for C_s into the previous equation, while taking the ratio of the frequency of the driving force to the natural frequency of the body on the mounting r, thus

$$r = \frac{\omega}{\omega_0}$$

(equation 10.7) giving

$$f_w = \left\{ \frac{F_0 \sqrt{[1 + (2\alpha/\omega_0)^2 r^2]}}{\sqrt{[(1 - r^2)^2 + (2\alpha/\omega_0)^2 r^2]}} \right\} \cos(\omega t - \gamma_s + \varphi)$$

$$= F_w \cos(\omega t - \gamma_s + \varphi). \qquad (11.11)$$

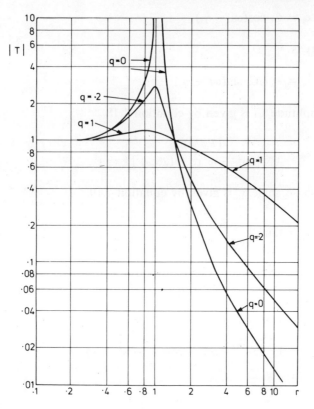

Figure 11.5 The absolute transmissibility function $|T|$ for the mounting of the machine in figure 11.4 as a function of frequency ratio $r\ (= f/f_0)$ and damping ratio $q\ (= b/b_c)$ where b_c is the damping coefficient for critical damping $(= 2\sqrt{(CM)}$ and f_0 is the frequency of free undamped vibrations of the system $[= (1/2\pi)\sqrt{(C/M)}]$.

The force on the wall is therefore sinusoidal, of the same frequency as the unbalanced force in the body, lags this force by the phase angle $(\gamma_s - \varphi)$ and has amplitude F_w. The ratio of the amplitude of the force on the wall to the amplitude of the driving unbalanced force is called the absolute transmissibility $|T|$, thus

$$|T| = \frac{F_w}{F_0}$$

$$= \left\{ \frac{\sqrt{[1 + (2\alpha/\omega_0)^2 r^2]}}{\sqrt{[(1 - r^2)^2 + (2\alpha/\omega_0)^2 r^2]}} \right\} \quad (11.12)$$

$$= \left\{ \frac{\sqrt{[1 + (\alpha/\pi f_0)^2 r^2]}}{\sqrt{[(1 - r^2)^2 + (\alpha/\pi f_0)^2 r^2]}} \right\} \qquad (11.13)$$

The expressions for the absolute transmissibility are often written in two other forms based on the following equivalences from chapter 3

$$\frac{\alpha}{\omega_0} = q$$

(equation 3.37) where q is the damping ratio of the mounting and

$$\frac{\alpha}{\omega_0} = \frac{1}{2Q},$$

(equation 3.53) where Q is the quality factor of the mounting. Then

$$|T| = \frac{\sqrt{[1 + (2qr)^2]}}{\sqrt{[(r^2 - 1)^2 + (2qr)^2]}} \qquad (11.14)$$

$$= \frac{\sqrt{[1 + (r/Q)^2]}}{\sqrt{[(r^2 - 1)^2 + (r/Q)^2]}} \qquad (11.15)$$

In figure 11.5 are graphs of the absolute transmissibility $|T|$ plotted against the normalised frequency r $(= f/f_0 = \omega/\omega_0)$ for a number of values of damping ratio q.

These graphs and the alternative forms of the equations describing them are central to the techniques of passive vibration isolation. The criterion for isolation must obviously be that the absolute transmissibility be as small as possible or

$$|T| \to 0 \qquad (11.16)$$

Inspection of the graphs of figure 11.5 shows that this requirement is met only at high values of r, that is, when the natural frequency of the body and its mounting f_0 is much less than the frequency of the driving force f. For a given mass M this means that the spring constant of the mounting C be small so that the mounting be very flexible. Conversely, for a given mounting of spring constant C it means that the mass M must be made large.

In the steady-state, the degree of damping incorporated in the mounting plays a contradictory role. The less damping there is (q small) at high

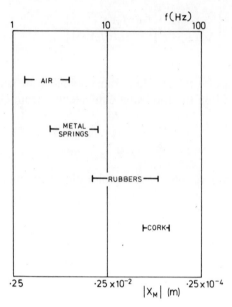

Figure 11.6 The ranges of operating frequencies and maximum displacements possible with different materials used for manufacturing machine mountings.

values of r, the lower is the value of $|T|$. Low damping then ensures low absolute transmissibility. However, if the source of the vibratory force goes through a start-up from rest, r rises from zero and $|T|$ undergoes the resonant behaviour near $r = 1$ before reaching the steady state. If the damping is very low, this resonance can be destructive. This can be avoided, or at least minimised, by incorporating some damping in the mounting. Usually, therefore, the choice of mounting is a compromise involving some damping to avoid excessive resonance near $r = 1$ and accepting the resulting slight increase in absolute transmissibility $|T|$ at the high operating frequency when r is large.

The types of materials used to construct mountings, the operating frequency ranges and the amplitudes of displacement ranges appropriate to these materials, are shown in figure 11.6. Air bearings have some damping and are suitable for low frequencies. Metal springs tend to have low damping as also does cork. Felt, useful for high frequencies, has a high damping performance. Rubber may be formulated with various additives to create a composite with any in a wide range of degrees of internal damping. Of course, discrete element dampers can often be used in conjunction with the spring mounting especially with metal leaf and coil springs.

VIBRATION ISOLATION AND CONTROL

Example 11.3

The force in the vertical direction generated by the flywheel of example 11.1 is 600 cos 52.4*t* N. Design a mounting to reduce the amplitude of the force on the floor to 0.01 that of the generated force. Also ensure that the maximum force on the floor is not greater than twice that generated.

Solution

In this problem, at ω_0, $|T| = 2$ and at ω, $|T| = 0.01$. Interpolating in figure 11.5, the curve for $q = 0.1$, would peak at $|T| = 2$ at about $r = 1$. So choose a damping ratio of 0.1. This curve would meet $|T| = 0.01$ at about $r = 20$. This is the operating point needed. When $r = 20$, $\omega = 52.4$ and so $\omega_0 = 2.62$ rad/s. But the mass $M = 10^3$ kg and therefore $C = 6.86$ kN/m. The damping coefficient for a damping ratio of 0.1 is then $b = 0.1 \times 2 \times \sqrt{(CM)} = 524$ kg/s.

11.5 Protection of Equipment against Vibrations

Sometimes an item of delicate equipment is to be installed in an environment, for instance on a bench, which is subjected to vibrations. It may not be possible or feasible to reduce these vibrations or to cut them off at the source(s) but it is possible to protect the equipment from the vibrations.

Consider the piece of equipment as a mass M standing on a mounting of spring constant C and damping coefficient b which in turn stands on the vibrating surface executing the motion $x_0(t)$. If $x_0(t)$ is harmonic then it is readily apparent that the absolute displacement transmissibility is identical with the absolute force transmissibility $|T|$ given in equations 11.12 to 11.15 and illustrated in figure 11.5 for a number of damping ratios. Thus the ratio of the displacement amplitudes is

$$\frac{|A_e|}{|A_0|} = |T| \tag{11.17}$$

$$= \frac{\sqrt{[1 + (2qr)^2]}}{\sqrt{[(r^2 - 1)^2 + (2qr)^2]}} \tag{11.18}$$

where A_e is the amplitude of the displacement of the equipment and A_0 is the amplitude of the displacement of the bench.

As in the case of vibration source isolation, the protection of the equipment requires the choice of mounting with some damping and with a natural frequency f_0 much less than the frequency present in the bench vibration. Materials from those mentioned in figure 11.6 may be selected for the mounting. In order to ensure that $|T|$ be as low as possible at the frequency of the ambient vibration, f_0 can be made small by choosing a mounting with a very flexible spring, low spring constant C, and also by increasing the effective mass M. Placing the equipment on a heavy (high mass) platform on top of the mounting thus reduces f_0, according to equation 2.12, and causes the absolute transmissibility $|T|$ to be much lower at that ambient vibration frequency.

11.6 Dynamic Vibration Absorbers

An alternative means of reducing unwanted vibrations is the dynamic vibration absorber. At its simplest this is a harmonic oscillator with mass M_1 and spring constant C_1 attached to the body of mass M which experiences the vibratory force $F_0 \cos \omega t$. This coupled arrangement is shown in figure 11.7. As a first approach assume no damping either in the original system or in the added dynamic absorber.

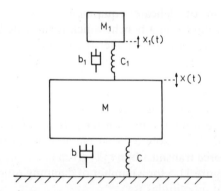

Figure 11.7 A dynamic vibration absorber, the harmonic oscillator $M_1/C_1/b_1$, associated with a vibrating system consisting of the mass/spring/damper $M/C/b$.

The new over-all system is a coupled system with two degrees of freedom x and x_1 like the system discussed in chapter 6. The force equations for the two masses, after equations 6.41 and 6.42, are

$$M\ddot{x} + Cx - C_1(x_1 - x) = F_0 \cos \omega t \qquad (11.19)$$

VIBRATION ISOLATION AND CONTROL

and

$$M_1\ddot{x}_1 + C_1(x_1 - x) = 0 \ . \tag{11.20}$$

Assuming that the two displacements x and x_1 follow the driving force $x = A \cos \omega t$ and $x_1 = A_1 \cos \omega t$, equations 11.19 and 11.20 become

$$(-\omega^2 - \omega_0^2)A - \omega_{01}^2(A_1 - A) = \frac{F_0}{M} \tag{11.21}$$

and

$$-\omega^2 A_1 + \omega_1^2 (A_1 - A) = 0 \ . \tag{11.22}$$

Here

$$\omega_0^2 = \frac{C}{M} \tag{11.23}$$

$$\omega_1^2 = \frac{C_1}{M_1} \tag{11.24}$$

and

$$\omega_{01}^2 = \frac{C_1}{M} \ . \tag{11.25}$$

Two simultaneous equations in A and A_1 emerge from equations 11.21 and 11.22 thus

$$A(\omega_0^2 + \omega_{01}^2 - \omega^2) - A_1 (\omega_{01}^2) = \frac{F_0}{M} \tag{11.26}$$

and

$$A(-\omega_1^2) + A_1(\omega_1^2 - \omega^2) = 0 \ . \tag{11.27}$$

If the ratio of the two masses be taken as g, where

$$g = \frac{M_1}{M} \tag{11.28}$$

the solutions to the simultaneous equations 11.26 and 11.27 may be shown to be

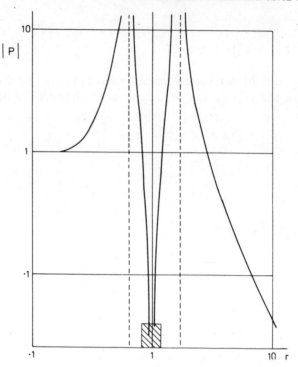

Figure 11.8 The absolute protection parameter $|P|$, the reduction in the displacement amplitude due to the addition of the dynamic absorber without damping, as a function of frequency ratio r. In a narrow range of r close to $r = 1$, $|P|$ is less than unity and so the vibration of the mass is reduced. However, on either side of this protection range, destructive resonances can occur.

$$A = \frac{(F_0/M)(\omega_1^2 - \omega^2)}{\omega_1^2\omega_0^2 - \omega_1^2\omega^2(1 + g) - \omega^2\omega_0^2 + \omega^4}$$

$$= \frac{(F_0/C)[(\omega_1/\omega_0)^2 - r^2]}{(\omega_1/\omega_0^2)[1 - (1 + g)r^2] - r^2 + r^4} \, . \quad (11.29)$$

$$A_1 = \frac{(F_0/C)(\omega_1/\omega_0)^2}{(\omega_1/\omega_0)^2[1 - (1 + g)r^2] - r^2 + r^4} \quad (11.30)$$

where $r = \omega/\omega_0$. In practical dynamic absorbers, M_1 and C_1 are chosen such that $\omega_1 = \omega_0$. Then equation 11.29 becomes

VIBRATION ISOLATION AND CONTROL

$$A = \frac{(F_0/C)(1 - r^2)}{1 - (2 + g) r^2 + r^4}.$$ (11.31)

The quantity $|A|/|F_0/C|$ is the ratio of the amplitude of the displacement of the body to be protected to the amplitude of the displacement of the same body when subjected to a static force F_0. It can be defined as a measure of the absolute protection $|P|$ provided by the dynamic vibration absorber.

$$|P| = \frac{|A|}{|F_0/C|}$$ (11.32)

and it is plotted against r in figure 11.8. $|P| < 1$ is the criterion for reduction of the vibratory motion of the main mass. From the figure, at $r = 1$, $|P| = 0$

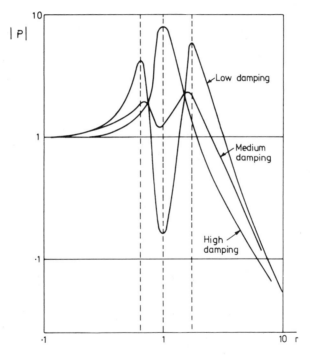

Figure 11.9 The introduction of damping into the dynamic absorber modifies the dependence of the absolute protection parameter $|P|$ on frequency ratio. The amount of protection is reduced in the protection range of frequencies, that is, $|P|$ is greater than in figure 11.8 but the resonances are also reduced. At very high damping in the absorber, its effectiveness as a protector against excessive vibrations of the machine is more and more reduced.

262   FUNDAMENTALS OF MECHANICAL VIBRATIONS

and so the main mass experiences no vibratory displacement. In the band of frequencies characterised by the cross-hatched range of values of r in the figure, $|P| < 1$ and some protection is afforded to the body. The absorber acts therefore as a band-stop filter over this range of frequencies — the so-called absorbing frequency band.

The incorporation of damping into the absorber results in the absolute protection index, $|P|$ curves in figure 11.9. Damping reduces the two resonances on either side of $r = 1$, a desirable feature in absorbers which have to deal with machines with start-up vibrations. As the amount of damping increases, the two resonances are eventually eliminated completely. Mild damping also, however, lowers the vibration reduction, increases $|P|$ in the vicinity of $r = 1$ and the more damping, the less protection is obtained.

Dynamic vibration absorbers with low damping are most appropriate in systems such as pneumatic drills, compacters, clippers, etc., which work at a constant speed and produce a fixed driving frequency. Such tools must often be held in the hand and, as will be seen in chapter 12, they should vibrate minimally for the comfort and safety of the operator.

Example 11.4

A truck is to be fitted for transporting delicate equipment by means of a dynamic vibration absorber. The mass of the truck when laden is 5 Mg and the effective spring constant of the suspension is 0.5 MN/m. Take the mass of the absorber to be 250 kg and find its spring constant and the range of frequencies for which it gives protection.

Solution

The resonant frequency of the truck may be calculated thus

$$f_0 = \frac{\omega_0}{2\pi} = \frac{10}{2\pi} = 1.6 \text{ Hz} .$$

In equation 11.28, $g = 0.05$. Taking $\omega_1 = \omega_0$, the spring constant of the absorber C_1 can be calculated, $C_1 = 25$ kN/m.

From equations 11.31 and 11.32, setting $|P| = 1$ yields the values of r between which $|P| < 1$. The value of $r > 1$, is found from equation 11.31

$$1 - r^2 = 1 - (2.05) r^2 + r^4$$

$$r = 1.02 \text{ or } f = 1.63 \text{ Hz} .$$

The value of $r < 1$ is found from

$$1 - r^2 = -r^4 - 1 + 2.05\, r^2$$

$$r = 0.98 \text{ and } f = 1.57 \text{ Hz}.$$

So the range of protection frequencies is from 1.57 to 1.63 Hz.

11.7 Experiments

(a) Measurement of Absolute Transmissibility

Absolute transmissibility is best measured with two accclerometers used simultaneously. One is used to measure the acceleration of the surface carrying the source vibration while the second measures the acceleration of the target or vulnerable body. The ratio of the amplitude of the latter to that of the former is by definition the absolute transmissibility. These measurements are made at frequencies across the spectrum so that a graph of $|T|$ versus frequency may be constructed.

If a particular mounting is to be assessed, a variable frequency vibrating source can be used as a more regular source and the vibration of the target body monitored with an accelerometer at each set frequency.

If a vibrometer is used as the vibration detector the resultant $|T|$ versus frequency graph should be identical to that found above.

(b) Vibration Reduction by Mounting Design

Given a vibrating surface, preferably that of a controllable vibrator, and an object of mass M, the aim is to choose a mounting to reduce the amplitude of the acceleration of the body when placed on the vibrating surface. The frequencies present in the ambient vibrations must first be measured and the spectrum characterised. Then for isolation, from figure 11.5, the natural frequency of the mass/mounting spring chosen must be much lower than the main frequencies in the vibration. This determines the range of values of C which will suffice. With some guidance from figure 11.6, a material for the spring in the mounting is chosen. Then the geometry of the springs must be decided — a pad, or three or four pads, their heights and cross-sectional areas. It may be necessary to measure the spring constant of the final spring. Finally, some damping may have to be incorporated to avoid resonant behaviour during stopping and starting. A damping ratio may be chosen from figure 11.5 and a discrete damper designed (see equation 3.15) to meet this requirement.

Finally the absolute transmissibility of the mounting must be measured as a function of frequency to confirm its performance as acceptable.

(c) Design and Performance of a Dynamic Vibration Absorber

Given an object of mass M on a spring of constant C that stands on a surface which can be vibrated at variable frequency, the object is to minimise the resonant vibration of the mass. A second mass M_1 attached to a spring of constant C_1 is attached to the first mass as in figure 11.7. Assume no damping coefficient b_1 initially. Choose M_1 and C_1 such that $C_1/M_1 = C/M$.

Now measure the absolute transmissibility of the vibration to M as a function of frequency. A graph along the lines of figure 11.8 should be obtained.

Then introduce a small amount of damping coefficient b_1 and repeat the measurement of absolute transmissibility to M at different frequencies. A graph such as in figure 11.9 should be found.

This procedure can be repeated for different values of b_1.

Problems

11.1 An electric motor of mass 200 kg runs at 1000 rotations per minute. It has a slight rotational imbalance. Design a mounting for the motor to ensure that the displacement of the floor is no more than 10^{-2} of that of the motor at full speed and no more than 3 at the worst case frequency.

11.2 If a rigid platform of 200 kg is first put under the motor of problem 11.1 what will the parameters of the mounting be to achieve the same performance as required.

11.3 A machine of mass 1 kg is subjected to an A.C. force of frequency 25 Hz. What spring constant and damping coefficient must be interposed between machine and bench to ensure that the bench experiences a maximum vibratory force of 0.05 of that acting on the machine?

11.4 A pneumatic drill of mass 40 kg has a principal frequency of 15 Hz. Design the parameters of a coupling between the drill and the operator to reduce the amplitude of the displacement of the operator's skin by a factor of 10.

11.5 If an engine generates vibrations of fundamental frequency 9 Hz with some third harmonic also and is placed on mounting pads, the

VIBRATION ISOLATION AND CONTROL

fundamental of the vibration transmitted to the floor is reduced by a factor of 5. The mass of the engine is 500 kg. What is the spring constant of the pads and by what factor is the third harmonic transmitted to the floor reduced? What is the damping coefficient?

11.6 A machine of mass 500 kg stands on tyres of spring constant 500 kN/m. It is observed to vibrate in the vertical direction with an amplitude of 3 mm and a period of 0.25 s. Find the expression for the driving force generated by the machine.

11.7 If the number of tyres under the machine of problem 11.6 is doubled, what will be the amplitude of the vibration?

11.8 An electric motor of mass 50 kg sits on rubber pads which deflect 5 mm under the static weight of the motor. It vibrates by an amplitude of 0.1 mm at full speed of 1500 revolutions per minute. What mass of sheet steel must be placed under the motor and over the pads to reduce the amplitude by a factor of 10?

11.9 If in each case of problem 11.8 a damper sufficient to achieve a damping ratio of 0.5 were added to the system, what would the two deflection amplitudes be?

11.10 A delicate instrument of 5 kg mass is to be transported in a carton mounted between two springs, vertically mounted, each of the same spring constant. The amplitude of motion of the instrument must be no more than 0.01 of that of the container at the steady running frequency of 3 Hz and must never exceed two times that of the container. What is the spring constant of each spring and what is the damping ratio of the mounting?

11.11 In a mounting without damping the amplitude resonance is threefold at 50 Hz, which is above the resonant frequency. At what frequencies would the amplitude transmitted be 0.1 and 0.01 of the source amplitude? If a damping ratio of 0.2 were introduced, at what frequencies would these amplitude reductions be achieved?

11.12 A sensitive instrument sits on a bench which has vibrations of 100 Hz. A cushion is placed under the instrument and reduces the displacements experienced by the instrument by a factor of 100. If the static displacement of the cushion is 1 cm and the damping coefficient of the cushion is 123 kg/s, find the mass and spring constant in the system.

11.13 A machine resonates at 40 Hz. A vibration absorber of mass 5 kg is to be used to provide protection to the machine. What is the spring constant of the absorber and over what range of frequency does it provide protection? The machine mass is 100 kg.

11.14 In a complex industrial set-up resonances can occur in the frequency range 5 to 20 Hz. If a series of 1 kg absorbers are to be used find how many such absorbers are needed, the spring constant of each one and the frequency range dealt with by each one. Take each value of g to be 0.1.

11.15 From equation 11.31 investigate the influence of the value of the mass ratio g on the width of the frequency range for which $|P| < 1$. Draw up a diagram with g on the vertical axis and the r values on the horizontal axis setting up the g-axis at $r = 1$.

Further Reading

Baker, J.K., *Vibration Isolation* (Oxford University Press, 1975)
Beranek, L.L., *Noise and Vibration Control* (McGraw-Hill, New York, 1971)
Bramer, T.P. *et al.*, *Basic Vibration Control* (Sound Research Laboratories, Sudbury, 1977)
Broch, J.T., *Mechanical Vibration and Shock Measurements* (Brüel & Kjaer, Naerum, 1973)
Harris, C.M. (Ed.), *Handbook of Noise Control* (McGraw-Hill, New York, 1957)
Petrusewicz, S.A. and Longmore, D.K. (Eds.), *Noise and Vibration Control for Industrialists* (Elek Science, London, 1974)

12 Vibrations and the Human Body

12.1 Introduction and Objectives

The aim in this chapter is to describe the effects of externally applied vibrations on the human body and to discuss the significance of two internally generated mechanical vibrations in the body.

After reading this chapter the student should be able to:

(a) describe the vibration sensors in the human body;
(b) discuss the physiological and pathological responses to vibration of the body;
(c) outline the schematic view of the body as an array of coupled mechanical oscillators;
(d) discuss the protection of the human body against over-exposure to vibrations;
(e) describe the genesis and significance of the ballistocardiogram (BCG);
(f) outline the origins and course of the mechanical wave travelling along the arteries in the body;
(g) discuss the medical diagnostic utility of the study of these biological vibrations.

12.1 Effects of Mechanical Vibrations on Human Beings

Many effects of mechanical vibrations on the human body have been described and investigated. The effects and their severity depend on the amplitude and frequency of the vibration and on the portions of the body experiencing the vibration.

The first such effect, and that with the lowest threshold, is simple perception or feeling. Such perception is mediated through many different detectors or sense receptors — nerve endings in the skin, pacinian corpuscles deeper in tissue, muscle spindles within muscles, Golgi tendon organs in muscles and the vestibular apparatus inside each side of the skull. As an example, the threshold for perception of tangential vibrations

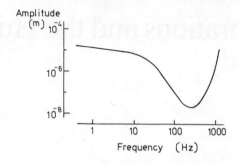

Figure 12.1 The threshold for perception of tangential vibrations applied to the finger-tips as a function of frequency. (After Harris).

applied to the fingertips is frequency-dependent and goes to a minimum of close to 10^{-8} m amplitude at 200 to 300 Hz as shown in figure 12.1.

The pacinian corpuscle is basically a nerve ending surrounded by a series of lamellae of elastic connective tissue similar in construction to an onion as illustrated in figure 12.2. As a mechanical system, the corpuscle may be represented roughly by the complex set of coupled mass/spring/damper oscillators along the lines shown in figure 12.3. The pacinian corpuscle is found experimentally to have a maximum response or minimum threshold in the range 100 to 300 Hz. This maximisation of the response is a resonance effect. The maximum transmissibility through the capsular mechanical system occurs within a specific resonant frequency range.

The perception of whole body vibrations also demonstrates frequency-dependent thresholds. Thus the thresholds for perception by a person standing and sitting on a platform vibrating vertically are shown in figure

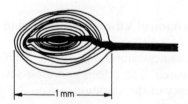

Figure 12.2 The lamellar structure of the pacinian corpuscle which is located within the skin, tendons and other tissues and can detect vibratory pressure. The many layers of elastic connective tissue surround a sensory nerve ending in which action-potentials are generated by the pressure applied to the exterior of the oval corpuscle.

VIBRATIONS AND THE HUMAN BODY   269

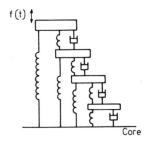

Figure 12.3 The mechanical properties of the elastic lamellar structure of the pacinian corpuscle may be roughly represented by a system of coupled mass/spring/damper sub-systems.

12.4. For this mode of vibration and where mere perception is the response, the minimum threshold is in the frequency range of 1 to 10 Hz.

Broadly, it is thought that for frequencies of vibration below 10 Hz the most sensitive detectors in the body are those in the vestibular apparatus. Above 10 Hz, deep receptors in the muscles take over while above 1 kHz the superficial skin receptors are most sensitive.

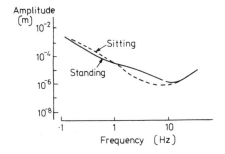

Figure 12.4 The threshold for perception of whole-body vertical vibrations applied to the sitting and standing human being as a function of frequency. (After Harris).

A vestibular apparatus is located within the temporal bone close to the inner ear on each side of the head. It contains two types of receptors the otolith organs — the utricle and saccule — which sense linear accelerations and the three semi-circular canals which detect angular accelerations. Figure 12.5 illustrates the shapes and relative positions of these structures. They are a set of interconnected chambers filled with liquid endolymph.

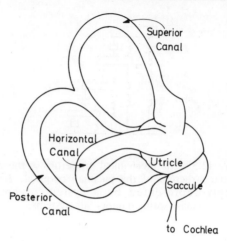

Figure 12.5 The general structure of a vestibular apparatus comprising the three, roughly perpendicular, semicircular canals and the two otolith organs, the utricle and saccule.

Each otolith organ has a patch of hair cells protruding into a flexible gelatinous membrane which in turn is surmounted by a carpet of dense calcium carbonate crystals, the otoconia. These crystals serve as an inertial mass which, when it experiences a tangential acceleration, elastically deforms the gelatinous membrane in shear and this stimulates the hair cells. The direction and sense of the acceleration are detected mainly through the sub-population of the hair cells stimulated. The action-potential signals from these hair cell nerve endings are carried to the brain via the eighth nerve. In the upright head the transducer in the utricle responds to horizontal linear acceleration of the head while that in the saccule detects vertical acceleration.

Each semi-circular canal in the upright head lies in one of the three perpendicular planes which are at roughly $\pi/4$ rad to the horizontal, saggital and coronal planes of the head. At the end of each canal, where it joins the utricle, is a swelling called the ampulla. Each ampulla is equipped with a patch of hair cells embedded in a flexible gelatinous flap called a cupula. The liquid endolymph in the canal acts as an inertial mass. When the head experiences angular acceleration in the plane of one of the semi-circular canals, the endolymph in that canal remains stationary while the cupula is bent and the hair cells stimulated. The sense of the angular acceleration, clockwise or anti-clockwise, is perceived by the brain from the sub-population of the hair cells active in any instance.

Beyond mere perception of a vibration, there also occur certain more complex psychosomatic responses to vibrations. Discomfort, annoyance, fatigue and impairment of task performance have been observed as responses to vibration. More objective consequences also happen. Resonance of the eyeball in its socket, which occurs in the vicinity of 30 Hz, blurs vision and reduces visual acuity. This can drastically impair the person's ability to perform many tasks. Nausea, vertigo and motion sickness are other ill-effects due to intermediate amplitude whole-body vibrations and mediated via the vestibular apparatus.

At still higher amplitudes of vibration, pathological effects such as damage to tissues and organs occur. One of the most widely studied such injuries, caused by long-term exposure of the fingers or toes to vibration amplitudes of 10^{-4} m or more, at frequencies around 100 Hz, is Raynaud's syndrome. The symptoms are pain, numbness and cyanosis of the fingers or toes. If irreversible, this injury can lead to gangrene and then amputation of the affected parts.

Other damaging effects of intense vibrations are internal tearing of tissues, internal haemorrhage and even death.

In order to understand some of these effects it is useful to consider the body as a complex of coupled mechanical oscillatory systems.

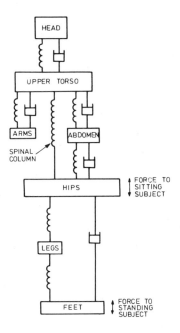

Figure 12.6 A broad representation of the whole body as a set of coupled mechanical vibratory systems.

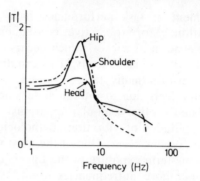

Figure 12.7 The absolute transmissibility $|T|$ of the acceleration applied to the feet of a standing person to the hip, shoulder and head as a function of frequency. Peaks indicate resonance. (After Harris).

12.3 The Body as a Mechanical System

As a mechanical system subjected to whole-body vibratory forces, the human body may be schematically represented by the pattern of coupled oscillatory systems shown in figure 12.6. A lumped discrete parameter model for the body is justified up to around 1 kHz because the wavelength of longitudinal waves in tissues at this frequency is greater than 1.5 m and is still more at lower frequencies. The various organs and tissues of the body therefore have dimensions much less than a wavelength.

Consider the situation of vertical vibratory forces applied to the person via a platform on which the subject is standing. One measure of the consequences of applying such a driving force is the absolute transmissibility (displacement or acceleration) from platform to various parts of the body. Figure 12.7 plots the absolute transmissibilities to hip, shoulder and head as functions of driving frequency. There is a resonance around 4 Hz related to the torso mass/leg spring constant system. There is a further resonance in the vicinity of 20 Hz related to the head mass/neck spring constant sub-system.

The natural or resonant frequencies of a number of body organs and sub-sections have been measured. Table 12.1 lists the ranges of resonant frequencies found for some of these sub-systems when the body is subjected to vertical vibrations.

Another quantitative measure of the response of the body to vibratory force is the driving point impedance Z defined in equations 4.22 and 4.34 as the ratio of the driving force to the resultant velocity at the point of

VIBRATIONS AND THE HUMAN BODY

Table 12.1 Resonant Frequency Ranges of Body Subsystems when Standing (Vertical Vibratory Force) (After Harris)

Body Organs	Frequency Range (Hz)
hips	2 to 8
arms/shoulders	2 to 8
chest	2 to 12
abdomen	2 to 14
lower back	6 to 12
head	8 to 27
bladder pressure	10 to 27
eyes/vision	12 to 27

application of the force. In general Z_m is complex and may be expressed thus

$$Z_m = |Z_m| e^{-j\varphi} . \tag{4.34}$$

Figure 12.8(a) and (b) are plots of $|Z_m|$ and φ respectively for vertical vibration of a standing person over a range of frequencies. At low

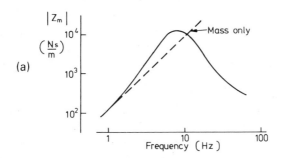

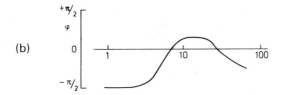

Figure 12.8 The driving point impedance (a) magnitude $|Z_m|$ and (b) phase angle φ of the standing human body as a function of frequency. The vibrations are applied vertically to the feet. (After Harris).

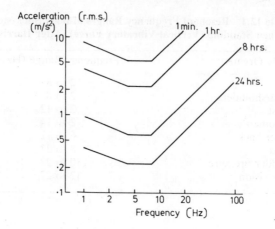

Figure 12.9 The recommended maximum vibratory accelerations to be applied vertically to the standing human being at different frequencies and durations of exposure. (From International Standards Organisation, after Broch and Beranek).

frequencies, the whole body behaves as a pure mass. As the frequency rises to the resonant frequency below 10 Hz, where $\varphi = 0$, the body acts as a damper or energy absorber. Then over a slightly higher band of frequencies the spring reactive element predominates but the spring constant itself is strongly dependent on frequency, rising as the frequency increases. The system is very non-linear. Then at still higher frequencies, the body acts as a net mass again ($\varphi < 0$).

The International Standards Organisation has promulgated maximum allowable vibration exposure criteria because of the dangers of destructive resonances and the associated ill-effects. Some of these recommendations are shown in figure 12.9. These graphs detail the maximum allowed root-mean-square vibrational accelerations for the standing whole body, over a range of frequencies, for various durations of exposure. They clearly isolate the frequency range from 4 to 8 Hz as the most vulnerable range. At lower and higher frequencies, either higher accelerations or longer exposure durations are tolerated by the body without untoward effects.

12.4 Ballistocardiography

During each contraction of the heart the pulsatile flow of blood out of the heart and around the approximately π rad bend of the aorta, as shown in figure 12.10, is an intrinsic source of vibrations in the body. Measurement

VIBRATIONS AND THE HUMAN BODY

Figure 12.10 The pulsed outflow of blood from the heart travels around the almost π rad curve of the aorta and thereby produces a pulsed force, mainly in the body axis direction, which accelerates various parts of the exterior of the body. These external accelerations are detected and recorded as the ballistocardiogram (BCG).

of the resultant vibratory motion of the surface of the body allows useful diagnostic insights into the dynamics of the heart.

The driving force is the repetitive pulsed change in the momentum of the blood as its direction is changed in coursing through the aorta. This force on the aorta is loosely coupled to the thorax which is coupled in turn to the head and legs as indicated in figure 12.6. The exact transmissibility between the accelerating mass of blood and any part of the body, such as the head or feet, can not be known nor can the exact force imparted to the blood be known but what can be measured and recorded is either the instantaneous acceleration (using an accelerometer) or the instantaneous displacement (using a vibrometer) of an external part of the body, along the axis of the body or in the saggital direction from left to right or in the coronal direction from front to back. These recorded signals constitute the ballistocardiogram (BCG).

A typical trace of a normal acceleration BCG, measured at the feet of a supine subject, is shown in figure 12.11. This pattern is repeated each heartbeat. For reference purposes the sequence of voltage deflections are labelled G, H, I, J, L, M and N waves.

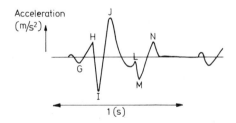

Figure 12.11 A recording of a typical period of a foot acceleration BCG in the foot-to-head axial direction. The lettered labelling is simply a convention for ease of reference.

If the feet to shoulder transmissibility indicated in figure 12.7 is approximately the same as for the blood mass in the torso to feet coupling of this case, then since the transmissibility is unity at low frequencies (< 4 Hz), the low frequencies present in the BCG of the figure 12.11 would be a good representation of the intrinsic blood acceleration and so of cardiac dynamics but the resonance effect in the vicinity of 8 to 10 Hz of figure 12.7, and the fall-off in the transmissibility at frequencies greater than 10 Hz, would indicate a poor correspondence between intrinsic blood dynamics and measured acceleration at those frequencies.

Of course a spectral analysis of the BCG, subsequently operated on by the transmissibility function, both magnitude and phase angle of figure 12.7, would allow a more accurate picture of the cardiac blood flow dynamics to be derived. Such signal processing could enhance diagnosis of cardiac diseases.

12.5 Pulsatile Wave Propagation along Arteries in the Body

An important class of vibrations within the body occurs in the walls of the arteries during normal pulsatile blood flow. A simplified view of an artery is a circular tube of cross-sectional area A and internal radius r with a thin flexible wall. Consider an elemental length dz of such a tube as shown in figure 12.12. During blood flow there is a pressure gradient $\partial p/\partial z$ along this segment and there is also a gradient of flow rate thus $- \partial \acute{q}/\partial z$. Here p is the instantaneous blood pressure and $\acute{q}$ is the volume flow rate per second passing a cross-section.

Continuity of flow demands that any net blood volume into the element expands the element. Therefore, each second

$$\acute{q} - (\acute{q} - d\acute{q}) = d\acute{q}$$

$$= \frac{\partial A}{\partial t} dz$$

$$= \frac{\partial A}{\partial p} \acute{p} \, dz$$

$$= \frac{1}{\acute{C}} \acute{p} \qquad (12.1)$$

where $1/\acute{C}$, is the compliance of the arterial wall in radial expansion, and

$$\frac{1}{\acute{C}} = \frac{\partial A}{\partial p} dz \; . \qquad (12.2)$$

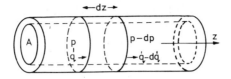

Figure 12.12 A segment of thin-walled tube, similar to an artery, along which the flow of liquid such as blood is occurring. The wall is distensible. Along the element of length dz there is a pressure drop dp and a decrement in flow rate dq́.

$1/C$, the compliance per metre tube length, a constant, characteristic of the wall, is defined thus

$$\frac{1}{C} = \frac{1}{C'dz}$$

$$= \frac{\partial A}{\partial p} \qquad (12.3)$$

and so equation 12.1, relating the flow rate gradient and the rate of change of pressure, becomes

$$\frac{\partial \acute{q}}{\partial z} = -\left(\frac{1}{C}\right)\acute{p} \qquad (12.4)$$

The motion of the blood along dz is controlled by the net force on the mass of blood $\rho A\,dz$, in dz, where ρ is the density of blood. This force arises from the pressure drop dp and is $A\,dp$. There is also a viscous resistance to this flow which may be found from Poiseuille's equation 3.10. The force equation, relating to the mass of liquid in the length dz is then

$$-A\,dp = \rho A\,dz\,\frac{\ddot{q}}{A} + \frac{8\eta A}{\pi r^4}dz\,\acute{q} \qquad (12.5)$$

where η is the coefficient of viscosity of the blood. Therefore

$$-dp = \frac{\rho dz}{A}\ddot{q} + \frac{8\eta\,dz}{\pi r^4}\acute{q} \qquad (12.6)$$

or

$$-\frac{\partial p}{\partial z} = \frac{\rho}{A}\ddot{q} + \frac{8\eta}{\pi r^4}\acute{q} \qquad (12.7)$$

Differentiating each side of this equation with respect to z produces

$$-\frac{\partial^2 p}{\partial z^2} = \frac{8\eta}{\pi r^4}\frac{\partial \dot{q}}{\partial z} + \frac{\rho}{\pi r^2}\frac{\partial \ddot{q}}{\partial z}$$

and from equation 12.4 this becomes

$$-\frac{\partial^2 p}{\partial z^2} = \frac{8\eta}{\pi r^4}\left(\frac{1}{C}\right)\dot{p} + \frac{\rho}{\pi r^2}\frac{\partial}{\partial t}\left(\frac{\partial \dot{q}}{\partial z}\right)$$

$$= \frac{8\eta}{C\pi r^4}\dot{p} + \frac{\rho}{C\pi r^2}\ddot{p} \,. \tag{12.8}$$

If the driving pressure is harmonic thus

$$p = P(z)\,e^{j\omega t} \tag{12.9}$$

equation 12.8 may be written

$$\frac{d^2 P(z)}{dz^2} = -j\omega\left(\frac{8\eta}{C\pi r^4}\right)P(z) + \left(\frac{\omega^2 \rho}{C\pi r^2}\right)P(z)$$

$$= \left(-j\frac{8\eta\omega}{C\pi r^4} + \frac{\omega^2 \rho}{C\pi r^2}\right)P(z) \,. \tag{12.10}$$

If this equation is written in the form

$$\frac{d^2 P(z)}{dz^2} = \gamma^2 P(z) \tag{12.11}$$

this is the wave equation similar to that of equation 7.20 describing the transverse wave on a stretched string except that γ is a complex propagation constant. If γ be written thus

$$\gamma = \beta - j\alpha \tag{12.12}$$

then

$$\gamma^2 = (\beta^2 - \alpha^2) - j2\alpha\beta$$

$$= \frac{\omega^2 \rho}{C\pi r^2} - j\frac{8\eta\omega}{C\pi r^4} \,. \tag{12.13}$$

From this relationship it can be shown that the real part of the propagation constant β is given by

$$\beta = \sqrt{\left\{\frac{\rho\omega^2}{2C\pi r^2}\left[1 + \sqrt{\left(1 + \frac{16\pi C\eta}{\omega^3\rho^2}\right)}\right]\right\}} . \quad (12.14)$$

The imaginary part of the propagation constant α is

$$\alpha = \frac{4\eta\omega}{C\pi r^4 \beta} \quad (12.15)$$

and is an attenuation constant for the wave.

The solution to the wave equation 12.11 then takes the form

$$p = (G\,e^{-j\beta z} + H\,e^{j\beta z})\,e^{-\alpha z}\,e^{j\omega t} \quad (12.16)$$

which represents a pair of waves each diminishing as it travels, one forward-moving and one reverse-moving. Each diminishes exponentially as it travels. H and G are amplitude parameters set by the initial conditions.

The speed of propagation of these waves

$$c = \frac{\omega}{\beta} \quad (12.17)$$

is a function of frequency and therefore the wave travelling along the tube is a dispersive wave.

The solution of equation 12.16 describes the pressure in the tube at any location z along the length and at any instant t. Consider a forward-moving wave only such that $H = 0$. From equation 12.3, the cross-sectional area of the tube at location z is related to the pressure thus

$$\frac{\partial A(z)}{\partial p} = \frac{1}{C}$$

and therefore

$$\begin{aligned}
A(z) &= \left(\frac{1}{C}\right)\int dp \\
&= -\left(\frac{1}{C}\right)p + A_0 \\
&= \left(\frac{1}{C}\right)G\,e^{-\alpha z}\,e^{-j\beta z}\,e^{j\omega t} + A_0 .
\end{aligned} \quad (12.18)$$

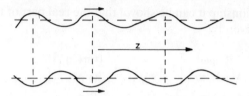

Figure 12.13 As the pressure wave and accompanying flow-rate wave travel forward along the tube there is also a bulging and contraction wave which propagates along the wall of the tube or artery.

The physical solution is the real part of $A(z)$ thus

$$\text{Re}[A(z)] = \left(\frac{G}{C}\right) e^{-\alpha z} \cos\left[\omega t - \beta z - \tan^{-1}\left(\frac{\beta}{\alpha}\right)\right] + A_0. \quad (12.19)$$

This is a gradually attenuating, travelling sinusoidal wave. It consists of alternating bulging and contracting of the cross-sectional area of the tube as shown in figure 12.13 with this disturbance propagating down along the tube.

The flow rate along the tube is also harmonic and, from equations 12.4 and 12.9, taking the forward-travelling wave only

$$\dot{q} = \left(\frac{1}{C}\right)\int \dot{p}\, dz$$

$$= \left(\frac{1}{C}\right)(j\omega)\int p\, dz$$

$$= -\left(\frac{\omega}{C(\beta - j\alpha)}\right) G\, e^{-\alpha z}\, e^{-j\beta z}\, e^{j\omega t} \quad (12.20)$$

The actual flow rate is the real part of this solution

$$\dot{q} = -\left[\frac{G\omega}{C\sqrt{(\alpha^2 + \beta^2)}}\right] e^{-\alpha z} \cos\left[\omega t - \beta z + \tan^{-1}\left(\frac{\alpha}{\beta}\right)\right]. \quad (12.21)$$

If the blood flow along the arterial tube obeys Poiseuille's equation then the spatial average blood velocity in the z-direction

$$\dot{z}_A = \frac{\dot{q}}{\pi r^2} \quad (12.22)$$

from equation 3.11. So this average blood velocity also is an attentuating travelling wave.

Doppler ultrasound and nucleotide scanning techniques permit the detection and measurement of some aspects of the travelling waves in the arteries of the intact human being. Valuable diagnostic data about the cardiovascular system may be derived from such measurements. For instance, arteriosclerosis, a very common and fatal disease of the arteries, is manifested as stiffening of the walls of the arterial tubes. The compliance $1/C$ is reduced by this disease. Inspection of equations 12.14 and 12.17 shows that the propagation speed in this disease is elevated above the normal. Furthermore, from equation 12.15 the attentuation or decay coefficient α also is reduced by this disease. Early detection of such changes and diagnosis of the degenerative changes allow corrective treatments to be used.

12.6 Experiments

(a) Measurement of Absolute Transmissibility in the Body

Measurement of the acceleration or displacement of a part of the body is difficult because of the scarcity of attachment locations for the transducers. The transducer must be attached as rigidly as possible to the surface to be examined. Yet even with this difficulty, some transmissibilities can be readily measured in the general laboratory. For instance the forearm can be driven at the elbow with a vibrator either horizontally, or vertically and the resultant acceleration of the fist measured. The accelerometer may also be fairly rigidly mounted on the clavicle and the transmissibility of the upper arm measured. If the foot is driven along the axis of the leg and the acceleration of the knee or hip is measured, the transmissibility of the leg can be determined. If a powerful vibrator is available to drive the whole body, standing or sitting, the transmissibilities in figure 12.7 may be determined.

(b) To Measure the Mechanical Impedance of the Body at Different Locations

If the harmonic force is applied as above to the part of the body of interest and the displacement of that point of application (measured with a vibrometer) is differentiated to yield the velocity, then the driving point impedance can be found. If the velocity of another location is measured then the transfer impedance can be calculated and plotted as a function of frequency.

(c) Measurement of the Ballistocardiograph

A crude way of obtaining the BCG is for the subject to lie supine with a bar tied across the two ankles. The accelerometer is attached to this bar and the amplified output of the transducer is recorded on a chart recorder. It is useful if it can be recorded together with an electrocardiograph (ECG). The BCG at the top of the head may also be recorded and the differences between this recording and the first one discussed in terms of the transmissibilities from the heart to the measurement points and in terms of the spectra of the two recordings. The effects of different postural changes on the records can be investigated. Such changes as bending the knees then raising the feet so that the legs below the knees are horizontal, lying on the right side and then on the left side, holding deep inspiration or deep expiration, etc. all could affect the BCG in various ways.

(d) Investigation of Arterial Blood Flow

The simple non-directional doppler ultrasound blood flow meter produces a voltage signal, the different frequencies of which are proportional to the velocities present in the blood. Frequency analysis of the signal yields the relative amplitudes of these frequency components and these amplitudes are roughly related to the volume of blood with the particular velocity. If such a measurement is made simultaneously at two locations along an artery, say 10 cm apart along the arm, and if the frequency analysis is carried out for each velocity signal, then a number of the parameters in equation 12.21 may be derived. c may be found from the delay time over the 10 cm for each frequency. The attenuation constant α may be found for each frequency from the reduction in amplitude over that length of flow. The relative values of $1/C$ the compliance per metre length, for different arteries may also be found by applying equation 12.21.

Problems

12.1 Convert the graphs of figure 12.4 into acceleration amplitude versus frequency.

12.2 Using figure 12.7, draw graphs (a) of the threshold amplitude of the hip versus frequency and (b) of the maximum allowable eight-hour hip acceleration (r.m.s.), from figure 12.9.

12.3 The BCG of a person of 150 kg yields the following spectral data: at 1 Hz, displacement amplitude, 10^{-3} m, acceleration amplitude, 40×10^{-3} m/s^2; at 2 Hz, displacement amplitude 2×10^{-4} m, acceleration amplitude,

32×10^{-3} m/s²; at 3 Hz, displacement amplitude, 10^{-5} m, acceleration amplitude, 4×10^{-3} m/s². Find the spring constant of the body/table mounting and the power output of the heart. Disregard damping.

12.4 A doppler ultrasound study of arterial blood flow along an artery in the leg yields the following data for two locations separated by 10 cm: relative amplitude of the velocity spectral component at 1 Hz, 0.8; transit time for this spectral component, 10 ms; radius of vessel 2 mm; density of blood, 10^3 kg/m³ and viscosity coefficient, 0.003 kg/m² s. Find the compliance per metre length of the arterial tube.

Further Reading

Beranek, L.L., *Noise and Vibration Control* (McGraw-Hill, New York, 1971)
Broch, J.T., *Mechanical Vibration and Shock Measurements* (Brüel & Kjaer, Naerum, 1973)
Dimarogonas, A.D., *Vibration Engineering* (West, St. Paul, 1976)
Harris, C.M. (Ed.), *Handbook of Noise Control* (McGraw-Hill, New York, 1957)
McDonald, D.A., *Blood Flow in Arteries,* 2nd edition (Arnold, London, 1974)
Noordergraaf, A., *Hemodynamics* in H.P. Schwan (Ed.) *Biological Engineering* (McGraw-Hill, New York, 1969)
Petrusewicz, S.A. and Longmore, D.K., (Eds.), *Noise and Vibration Control for Industrialists* (Elek Science, London, 1974)
Roberts, C. (Ed.) *Blood Flow Measurement* (Sector, London, 1972)
Starr, I. and Noordergraaf, A., *Ballistocardiography in Cardiovascular Research* (North Holland, Amsterdam, 1967)
Tempest, W. (Ed.) *Infrasound and Low Frequency Vibration* (Academic, London, 1976)
Weissler, A.M., (Ed.), *Noninvasive Cardiology* (Grune & Stratton, New York, 1974)
Woodcock, J., (Ed.) *Clinical Blood Flow Measurement* (Sector, London, 1976)
Woodcock, J., *Theory and Practice of Blood Flow Measurement* (Butterworths, London, 1975)

Answers to Problems

Chapter 1

1.1 142 N at 15.1° to x axis
1.2 267 N at 39.7° to vertical and 347 N at 79.4° to vertical
1.3 92.2 N
1.4 6.9°
1.5 19.6 N, 2 J
1.6 1000 kg, 50 m
1.7 698.1 N, 40 km, 2 km/s
1.8 1.11 s, 11.1 m
1.9 31.4 kN, 41.7 turns
1.10 98.2 MN/m
1.11 8.66 mm/s
1.12 3.75 MN/m
1.13 785 TN/m^2
1.14 0.57 MW
1.15 3.14 TN/m^2, 12.5 mm
1.16 2.6 rot/s
1.17 2.79 kN/m, 2.19 MJ

Chapter 2

2.1 0.32 Hz, 3.14 s
2.2 5 cos($2t - 1.53$)
2.3 790 N/m, 62 mm, + 1996 kg
2.4 19.6 kN/m, 2.23 Hz, 0.72 cos($14t - 1.47$) m
2.5 0.04 cos($158t - 0.31$) m
2.6 ±0.02 cos($158t$) m, ± 100 cos($158t$) N
2.7 3.98 GN/m^2
2.8 0.88 mN m/rad
2.10 3.37 s
2.11 2.48 Mg
2.12 9.05 kN/m
2.13 11.2 m, 35.1 m/s, 2.3 s
2.15 2.46 s
2.16 53.3 kN/m, 13.3 kN/m

ANSWERS TO PROBLEMS

Chapter 3

3.1 0.55, 28°
3.2 5.05 kg/s, 35.7 $e^{-0.25t}$ N
3.3 61.3 N, wall: 0.58, ground: 0.25
3.4 0.13 N
3.5 80
3.6 2 kg/s, 0.23 $e^{-5t}\cos(8.7t - 0.5)$ m
3.7 0.2 − 0.2 $e^{-5t}\cos(8.7t)$ m
3.8 24 kg/s, 1.2 Hz
3.9 0.174 + 0.28 $e^{-0.75t}\cos(7.5t + 2.4)$ m
3.10 0.1 + 0.01 $e^{-2t}\cos(9.8t - 0.2)$ m
3.12 0.33, 0.06
3.13 32.7 N m s/rad, 16 s
3.14 0.39 MN/m, 3.8 Mg/s, 0.3, 0.87
3.17 7.6 × 10^3 kg/s

Chapter 4

4.1 0.32 Hz, 1, 0.16 Hz and 0.48 Hz
4.2 11.08, 1.44 rad
4.3 0.41 m/s, 1.24 m/s, 0.42 m/s
4.4 0.24 1/s, 2.4 kg/s, 6.25, 43.4 cos (10t) N
4.5 1.08 W, 391.5 W
4.7 $\dfrac{\omega_0^2}{\sqrt{(\omega_0^2 - 2\alpha^2)}}$; $\dfrac{\omega_0^2 G_0}{2\alpha\sqrt{(\omega_0^2 - \alpha^2)}}$

4.8 0.77, 14.9, 1.23 rad, 0.067 m, 0.067 m/s, 11.3 mW
4.9 0.1 W, 0.2 m/s, 0.052 m
4.10 0.2 Ω, 1.59 MHz, 250 W, 50 A
4.12 0.51 cos (6.28t + 0.02) mm
4.13 2 mm, 0.8 mm, 0.06 μm, 1.41, 56.3 Hz, 36.4 Hz and 76.2 Hz

Chapter 5

5.1 (a) $\dfrac{1}{\pi}\left(1 + \dfrac{\pi}{2}\cos t + \dfrac{2}{3}\cos 2t - \dfrac{2}{15}\cos 4t...\right)$

(b) $\dfrac{2}{\pi}\left(1 + \dfrac{4}{3}\cos 20t - \dfrac{4}{15}\cos 40t + \dfrac{4}{35}\cos 60t...\right)$

(c) $\dfrac{2}{\pi}\left[\sin\left(\dfrac{2\pi}{T}t\right) + \dfrac{1}{3}\sin\left(\dfrac{6\pi}{T}t\right) + \dfrac{1}{5}\sin\left(\dfrac{10\pi}{T}t\right) + \dfrac{1}{7}\sin\left(\dfrac{14\pi}{T}t\right)\cdots\right]$

(d) $\dfrac{kT}{\pi}\left[\sin\left(\dfrac{2\pi}{T}t\right) - \dfrac{1}{2}\sin\left(\dfrac{4\pi}{T}t\right) + \dfrac{1}{3}\sin\left(\dfrac{6\pi}{T}t\right) - \dfrac{1}{4}\sin\left(\dfrac{8\pi}{T}t\right)\cdots\right]$

(e)

$\dfrac{2}{T\omega_c}\sin\dfrac{\omega_c\tau}{2} + \dfrac{2}{T}\left[\dfrac{\sin\left(\dfrac{2\pi}{T} - \omega_c\right)\dfrac{\tau}{2}}{\dfrac{2\pi}{T} - \omega_c} + \dfrac{\sin\left(\dfrac{2\pi}{T} + \omega_c\right)\dfrac{\tau}{2}}{\dfrac{2\pi}{T} + \omega_c}\right]\cos\left(\dfrac{2\pi}{T}t\right)$

$+ \dfrac{2}{T}\left[\dfrac{\sin\left(\dfrac{4\pi}{T} - \omega_c\right)\dfrac{\tau}{2}}{\dfrac{4\pi}{T} - \omega_c} + \dfrac{\sin\left(\dfrac{4\pi}{T} + \omega_c\right)\dfrac{\tau}{2}}{\dfrac{4\pi}{T} + \omega_c}\right]\cos\left(\dfrac{4\pi}{T}t\right)$

$+ \dfrac{2}{T}\left[\dfrac{\sin\left(\dfrac{6\pi}{T} - \omega_c\right)\dfrac{\tau}{2}}{\dfrac{6\pi}{T} - \omega_c} + \dfrac{\sin\left(\dfrac{6\pi}{T} + \omega_c\right)\dfrac{\tau}{2}}{\dfrac{6\pi}{T} + \omega_c}\right]\cos\left(\dfrac{6\pi}{T}t\right)\cdots$

5.2 (a) $\dfrac{\pi}{2}\left(\dfrac{\sin(1+\omega)\dfrac{\pi}{2}}{(1+\omega)\dfrac{\pi}{2}} + \dfrac{\sin(1-\omega)\dfrac{\pi}{2}}{(1-\omega)\dfrac{\pi}{2}}\right)$

(b) $\dfrac{\pi}{20}\left(\dfrac{\sin(10+\omega)\dfrac{\pi}{20}}{(10+\omega)\dfrac{\pi}{20}} + \dfrac{\sin(10-\omega)\dfrac{\pi}{20}}{(10-\omega)\dfrac{\pi}{20}}\right)$

(c) $-T\left(\dfrac{\sin\omega T}{\omega T}\right) - j\omega T^2\left(\dfrac{\sin\dfrac{\omega T}{2}}{\dfrac{\omega T}{2}}\right)^2$

(d) $j\dfrac{2}{\omega^2}\left(\cos\dfrac{\omega T}{2} - \sin\dfrac{\omega T}{2}\right)$

ANSWERS TO PROBLEMS

(e) $\dfrac{\tau}{2}\left(\dfrac{\sin(\omega_c+\omega)\dfrac{\tau}{2}}{(\omega_c+\omega)\dfrac{\tau}{2}}+\dfrac{\sin(\omega_c-\omega)\dfrac{\tau}{2}}{(\omega_c-\omega)\dfrac{\tau}{2}}\right)$

5.3 (a) $U(t) - e^{-at}$
(b) $U(t) + t^2$
(c) $U(t) + \cos at$
(d) $U(t) - \cos at$
(e) $U(t) + e^{-at} - \dfrac{a}{2} t e^{-at} - \dfrac{a^2}{2} t^2 e^{-at}$

5.4 $-0.31\, e^{-0.67t} \cos(1.77t + 1.47)$ m

5.5 $-0.49\, U(t) + 0.025 \cos 4.46t$ m

5.6 (a) e^{-t}
(b) $U(t) - e^{-t}$

(c) $\left(\dfrac{1}{1+\omega^2}\right)[\delta(t) - e^{-t}]$
(d) $k(-U(t) + e^{-t} + t)$

5.7 (a) $e^{-t} - 0.5t\, e^{-t}$
(b) $0.5t\, e^{-t}$

(c) $\dfrac{2\omega^2}{(\omega^2+1)^2}\cos \omega t + \dfrac{\omega(\omega^2-1)}{(\omega^2+1)^2}\sin \omega t + \dfrac{1}{2(\omega^2+1)}t\, e^{-t}$
$- \dfrac{2\omega^2}{(\omega^2+1)^2} e^{-2t}$

(d) $k(U(t) - e^{-t} - 0.5\, t\, e^{-2t})$

5.8 (a) $-e^{-t} + 2e^{-2t}$
(b) $e^{-t} - e^{-2t}$

(c) $\dfrac{3\omega^2}{(1+\omega^2)(4+\omega^2)}\cos \omega t + \dfrac{\omega(1-2\omega^2)}{(1+\omega^2)(4+\omega^2)}\sin \omega t$
$+ \dfrac{1}{(1+\omega^2)} e^{-t} - \dfrac{4}{(4+\omega^2)} e^{-2t}$

(d) $\dfrac{k}{2}[U(t) - 2\, e^{-t} + e^{-2t}]$

Chapter 6

6.1 0.36 Hz, 0.39 Hz, 0.5 cos(2.24t − 1.47) m,
0.5 cos (2.45t − 1.46) m, 0.25 [cos(2.24t − 1.47) + cos (2.45t − 1.46)]
0.25 [cos(2.24t − 1.47) − cos(2.45t − 1.46)]

6.2 6.9 kg/s

6.3 1.18 Hz, 8.01 Hz

6.4 body: − 10 cos 8.2t − 0.005 cos 50.2t mm,
chassis: − 8 cos 8.2t + 0.2 cos 50.2t mm

6.5 11.8 m/s, 80.1 m/s

6.6 0.5 Hz, 0.85 Hz

6.7 − 5(cos 3.16t + cos 5.37t) mm, − 5(cos 3.16t − cos 5.37t) mm

6.8 63.3 kg/s, 3, 21.1 kg/s, 0.33

6.9 $\dfrac{1}{2\pi}\sqrt{\left[\dfrac{C(2M_1 + M_2)}{M_1 M_2}\right]}, \dfrac{1}{2\pi}\sqrt{\left(\dfrac{C}{M_1}\right)}$

6.11 $\dfrac{1}{2}\left(\dfrac{w}{L}\right)^2 C Q_2^2 \cos^2(\omega^2 t + \gamma_2)$

Chapter 7

7.1 10^4 N

7.2 0.8 Hz, 1.25 m/s, 1.56 m, 0.005 m/s, 0.025 m/s^2

7.3 1.06 s, 6.73, 46.2 e$^{-j1.46}$, 0.15 m/s, 0.12 W

7.4 2.32 s, 2.32 s, 0.03

7.5 22.4 m/s, 0.50 N

7.6 160 m/s, 1280 N, 49.4 cos(638t + γ) m/s

7.7 0.04 cos(5t − 2z − 0.52) m, 0.80 Hz, 2.5 m/s

7.8 1081 MN/m^2

7.9 $\bar{\bar{y}} = \dfrac{\pi \sqrt{(A^2 + B^2)}}{L} c$

7.10 10 Hz, 2.72 sin(62.8t − γ) N

7.11 $C = -A,\ D = -B,\ \dfrac{\omega L}{2c} = \dfrac{\pi}{4}(1.19, 2.99, 5, 7...)$

7.12 26.9 MN/m^2, 0.2 s

ANSWERS TO PROBLEMS

7.13 $c_t^2 = \omega\sqrt{\left(\dfrac{I}{A}\right)c_1}$

7.14 $\sqrt{\left[\dfrac{\omega}{2}\sqrt{\left(\dfrac{EI}{\rho A}\right)}\right]}$

7.15 0.68 m/s, 5.1 m/s, 10.40 Hz, 5.1 Hz
7.16 0.68 m/s, 5.06 m/s, 10.40 Hz, 5.06 Hz
7.17 28.7 Hz, 53.9 Hz
7.18 1.67 m

7.19 7.04ω (spring), $\dfrac{1.42}{\sin(0.01\omega)}$

7.20 0.17 m, 458 Hz, 717.8 Hz
7.21 0.369 m, 458 Hz, 717.8 Hz
7.22 4.59 kN/m
7.23 332.3 Hz, 44
7.24 338.4 Hz, 43
7.25 (a) 0.247 m (b) 0.290 m
7.26 0.63 GN/m^2

Chapter 8

8.1 Q: 5.7 km/s, BT: 5.7 km/s, PZT-4: 3.1 km/s, PZT-51: 2.7 km/s
8.2 Q and BT: (a) 28.5 m, (b) 0.285 m, (c) 2.85 mm; PZT-4: (a) 15.5 m, (b) 0.155 m, (c) 1.55 mm; PZT-51: (a) 13.5 m, (b) 0.135 m, (c) 1.35 mm
8.3 (a) Q: 238 V, BT: 79 V, PZT-4: 94 V, PZT-51: 97 V; (b) Q: 6.7×10^{-9} m, BT: 97.8×10^{-9} m, PZT-4: 152.7×10^{-9} m, PZT-51: 248.9×10^{-9} m
8.4 Q: 5.7 MHz, BT: 5.7 MHz, PZT-4: 3.1 MHz, PZT-51: 2.7 MHz
8.5 Q: 42 N, BT: 12.6 N, PZT-4: 106 N, PZT-51: 103 N
8.6 Q: 0.25×10^6 V, BT: 0.26×10^6 V, PZT-4: 0.17×10^6 V, PZT-51: 0.10×10^6 V
8.7 Q: 2×10^{-10} C, BT: 1.9×10^{-8} C, PZT-4: 2.9×10^{-8} C, PZT-51: 4.8×10^{-8} C
8.8 Q: 17.5 kPa, BT: 3.4 MPa, PZT-4: 2.2 MPa, PZT-51: 2.7 MPa

Chapter 9

9.1 $0.01 (1 + \cos 10t)$ m $\quad 0 \text{ s} \leq t \leq 0.157 \text{ s}$
$0.01 [1 - 10(t - 0.157)] \quad 0.157 \text{ s} \leq t \leq 0.357 \text{ s}$
$-0.01 [1 + \cos 10(t + 0.357)] \quad 0.357 \text{ s} \leq t \leq 0.671 \text{ s}$
$-0.01 [1 - 10(t - 0.671)] \quad 0.671 \text{ s} \leq t \leq 0.871 \text{ s}$
$0.01 [1 + \cos 10(t + 0.871)] \quad 0.871 \text{ s} \leq t \leq 1.028 \text{ s}$

9.2 $5 \times 10^3 (1 + 25x^2) x$ N, $0.0203 \cos 6.1t - 0.000006 \cos 18.2t$ m

9.3 $- 0.816 \cos 50t$ mm

9.4 $0.02 \sin 7.07t$ m $\quad 0 \text{ s} \leq t \leq 0.074 \text{ s}$
$0.02 \sin 7.07(t + 0.296) \quad 0.074 \text{ s} \leq t \leq 0.222 \text{ s}$
$0.02 \sin 7.07(t + 0.592) \quad 0.222 \text{ s} \leq t \leq 0.296 \text{ s}$

9.5 $9.9(1 + 100x^2) x$ kN

9.7 20.1 N/m, $0.010 \, e^{-0.08t} \cos 12.56t - 0.001 \, e^{-0.8t} \cos 1.56t$ m

9.9 $0.0008 \cos 6t + 0.00001 \sin 6t$ m

9.10 2.02 Hz, 1.98 Hz and 2.04 Hz, 33.7

9.11 $0.254 \, e^{-0.52t} \cos 0.66t - 0.154 \, e^{-0.86t} \cos 0.4t$ m

9.12 0.13 Hz, 0.056 Hz and 0.295 Hz

9.13 9.86 N/m, 0.0025, $\pm$ 0.0025 m, 4

9.14 0.0023

9.15 $(0.01 - 0.000025) \cos 7.53t + 0.000025$ m $0 \text{ s} \leq t \leq \dfrac{T}{4}$ s

$- (0.01 - 0.000075) \cos 7.53t - 0.000025 \quad \dfrac{T}{4} \leq t \leq \dfrac{T}{2}$

$(0.01 - 0.000125) \cos 7.53t + 0.000025 \quad \dfrac{T}{2} \leq t \leq \dfrac{3T}{4}$

$- (0.01 - 0.000175) \cos 7.53t - 0.000025$ m $\dfrac{3T}{4} \text{ s} \leq t \leq T$ s

Chapter 10

10.1 0.084 kg, 0.38 s

10.2 $- 14.8 \cos 45t - 3.0 \cos 9t$ m/s^2, $0.007 \cos 45t + 0.0004 \cos 90t$ m

10.3 15.9 Hz, 35 Hz

10.4 4.8 mm

10.5 14.7 Hz

10.6 (a) 11.1, 7.9, 1.57 s, + 0.47 rad
(b) 0.051, 0.036, 6.28 s, $-$ 0.2 rad
(c) 70.7, 50, 0.126 s, + 0.79 rad
(d) 11.4, 8.1, 0.032 s, + 0.91 rad

ANSWERS TO PROBLEMS

10.7 $1.27\, e^{-5(x-0.71)^2}$

10.8 $\bar{x} = 0.71,\ \sigma^2 = 0.1$

10.9 $\dfrac{\cos 2(\tau + T)}{2} = R(\tau),\ 0.5$

10.10 $R(h) = 1 - \dfrac{2}{\tau}(|h| - nT),\ nT \leq |h| \leq nT + \dfrac{\tau}{2}\ (n = 0,1,2\ldots)$

$\qquad\qquad = 0 \qquad\qquad\qquad\text{all other } h$

10.11 $R(h) = 1 - \dfrac{4}{\tau}(|H| - nT),\ nT \leq |h| \leq nT + \dfrac{T}{4}\ (n = 0,1,2,3\ldots)$

$\qquad\qquad = 0 \qquad\qquad\qquad\text{all other } h$

10.12 $f_m \left(\dfrac{\sin \pi f_m \tau}{\pi f_m \tau} \right)^2,\ \sigma^2 = f_m$

10.13 0

10.14 $100\delta(\tau)$

Chapter 11

11.1 3.5 kN/m, 335 kg/s
11.2 7 kN/m, 670 kg/s
11.3 247 N/m, 6.3 kg/s
11.4 14.1 kN/m, 300 kg/s
11.5 9.4 kN/m, 0.05, 866 kg/s
11.6 554 cos 25.1t N
11.7 0.81 mm
11.8 408 kg
11.9 0.1 mm, 0.01 mm
11.10 2.2 N/m, 0.1
11.11 125 Hz, 542 Hz, 229 Hz, 1043 Hz
11.12 20 kg, 19.6 kN/m
11.13 316 kN/m, 39.2 Hz to 41 Hz
11.14 1.18 kN/m: 5.00 to 5.74 Hz
   1.55 kN/m: 5.74 to 6.59 Hz
   2.15 kN/m: 6.59 to 7.56 Hz
   2.69 kN/m: 7.56 to 8.68 Hz
   3.56 kN/m: 8.68 to 9.96 Hz
   4.68 kN/m: 9.96 to 11.43 Hz

6.16 kN/m: 11.43 to 13.12 Hz
8.12 kN/m: 13.12 to 15.06 Hz
10.70 kN/m: 15.06 to 17.28 Hz
14.09 kN/m: 17.28 to 19.83 Hz
18.54 kN/m: 19.83 to 22.76 Hz

Chapter 12

12.3   60 kN/m, 358 mW
12.4 0.092×10^{-12} 1/N

Index

Absorber, dynamic 258, 264
Acceleration
    angular 6, 17, 147, 157
    linear 2, 17, 38, 45
Accelerometer 94, 225, 229, 244, 263, 275, 282
Action potential 268
Air column 168
Analogue, electrical 37, 64, 92, 139, 171
Analyser, frequency or spectrum 124, 235, 241
Analysis, harmonic 97, 117, 124, 201, 234
Antinode 153, 168, 176
Artery 276
Attenuation constant 279
Autocorrelation function 238
Axis
    neutral 11, 154
    of symmetry 154

Balancing 250
Ballistocardiography (BCG) 274, 282
Bandwidth 241
Barium titanate 188
Beam
    cantilever 11
    three-point loading 11, 175, 249
Bessel functions 166
Blood flow 276, 282
Boundary conditions 151, 158, 161, 166, 170, 191

Cantilever (see beam)
Capacitance, electrical 37, 64, 92, 139, 173
Charge, electrical 37, 174
Coefficient of
    damping 38, 46, 51, 72, 89, 127, 174, 205, 222, 252
    decay 52, 65, 87
    friction, dynamic 44, 46
    friction, static 45, 66, 214
    viscosity 48, 67
Compliance
    mechanical 82, 182
    of artery wall 276, 282
Compressibility 82, 182
Computer
    analogue 39, 41, 65, 68, 93
    digital 225
Convolution 114, 122
Coupling coefficient 187
Criteria, exposure 274
Current, electrical 16, 37, 139, 174, 224

Damper 49
Damping (see also overdamping and underdamping)
    coefficient (see coefficient of damping)
    critical 54, 76
    ratio 56, 63, 74, 255, 263
    static friction 214, 218
Decay constant or coefficient (see coefficient of decay)
Decrement, logarithmic 62, 65
Delta function 112, 116, 120
Density 157, 161, 169, 174, 188
    of blood 277
Diagram
    free-body 24, 33, 36, 51, 73, 128, 137
    phase plane 27, 53
Dielectric
    constant 183
    displacement 183
Dirac delta function (see delta function)
Displacement
    angular 5, 231
    linear 2, 38, 149, 159, 174, 225, 231
Distribution function 236

Doppler ultrasound 281
Duffing equation 204
Dynamic vibration absorber (see absorber)

Earthquake 224
Eccentricity 248
Elasticity 10
Electric field strength 183
Electrode 181
Energy
　conservation of 15, 29
　internal 183
　kinetic 14, 28, 47, 59, 145
　mechanical 14, 17, 28, 48, 59, 134, 183
　potential 15, 28, 59, 145
Enthalpy 183
Entropy 183
Equilibrium
　of forces 4
　static 157
Exact differential 183
Exposure criteria for vibration (see criteria)

Fatigue 223
Field strength 183
Filter, band-stop 262
Fletcher's trolley 18
Flywheel 19
Force 2, 17, 38, 44, 72, 204, 211, 252
　centrifugal 248
　compressive 8, 156
　shear 158
　tensile 8, 149, 155, 163, 174
　transducer 232
Fourier analysis (see analysis)
Fourier series 97, 117, 124, 201, 234
Fourier transform or Fourier integral 110, 241
Fracture 12, 223
Free-body diagram (see diagram)
Frequency 17, 25
　angular 24, 52, 65
　fundamental 98, 152, 159, 162, 167, 201, 206
　harmonic 98, 152, 162, 170, 201, 206
Friction
　coefficient of dynamic (see coefficient)
　coefficient of static (see coefficient)
　solid-body 44
　static, damping 214, 218
Function
　Bessel (see Bessel functions)
　causal 112
　delta (see delta function)
　distribution (see distribution function)
　even 100
　Gaussian (see Gaussian distribution)
　odd 102, 105
　spectral density 241
　step 105
　system transfer 119

Gas constant, universal 169
Gaussian distribution 237
Gibbs'
　free energy 184
　phenomenon 106
Golgi tendon organ 267

Half-power points or 3 dB points 87, 212
Harmonics (see frequency)
Heat 47, 59
Hooke's law 8, 12, 19, 155, 182

Impedance, mechanical 17, 79, 85, 93, 272, 281
　complex 80
Inductance, electrical 37, 64, 92, 139, 171, 174
Initial conditions 25, 30, 114, 118, 130, 151, 158, 162, 167, 191, 215
Instability 205
Isolation of vibration 252, 257

Kundt's tube 175

Laplace transform 110
Laplacian operator 164
Law
　Hooke's (see Hooke's law)
　Newton's second (see Newton's second law)
　Pouiseuille's (see Poiseuille's law)

INDEX

Lead zirconate titanate 188
Least squares error 104
Linearity 113, 116

Mass 2, 16, 23, 38, 44, 51, 72, 127, 145, 147, 201, 214, 222, 227, 248, 252
  molecular 169
Membrane 163
  drum 176
Mode co-ordinates 135
Modes 127
Modulus
  bulk 169
  shear 12, 40
  Young's 10, 20, 155, 160, 174, 182, 232
Moment
  bending 5, 17, 147, 156
  turning 5, 17, 147, 156
Moment of inertia
  area 11, 40, 156
  mass 7, 19, 147
Momentum
  angular 6
  linear 3, 17
Motion, simple harmonic (S.H.M.) 24
Motor effect 181
Mounting materials 256

Nerve spindle 267
Neutral axis (see axis)
Newton's second law 2, 6, 18, 45, 157
Node 153, 168, 175, 176
Noise 224
Normal co-ordinates 135
Nucleotide scanning 281

Orthogonality 98
Oscillator
  coupled 127
  damped 51, 67, 72
  harmonic 23, 35, 40, 146, 218
  rotational 34
Oscilloscope 38, 40, 68, 142, 175, 225, 245
Overdamping 55, 76 (also see damping)
  hysteretic 208
Overtones 153, 159, 167

Pacinian corpuscle 267
Pendulum
  compound 147, 174
  simple 32, 40, 67, 217
  torsion 40, 68
Period 25, 38, 97, 107, 148, 178, 203, 218
Phase plane diagram (see diagram)
Piezoelectric
  constant 183, 186, 187
  crystal 182, 227
  effect 180
  relationships 182, 185, 197
Plasticity 12
Poiseuille's law 49, 67, 277
Potential difference 182, 196 (also see voltage)
  divider 231
Power 16, 17, 83
  factor 85, 93
  instantaneous 83
  resonance 87, 211
Pressure 17, 169
Preventive maintenance 224
Probability density 235
Propagation
  constant 151, 165, 278
  speed (see speed of propagation)
Protection
  against vibration 257
  parameter 260
Pulse 107, 244

Q value or quality factor 63, 65, 89, 91, 94, 255
Quartz 188

Radius of curvature 154
Random vibrations 235
Ratio, damping (see damping)
Raynaud's syndrome 271
Relaxation time 46, 53, 65
Resistance, electrical 38, 64, 92, 139, 174
Resonance
  displacement 76, 136, 175, 194, 211, 256, 259, 272
  power 87, 211
  tube 175
  width 89

Response
  impulse 119
  psychosomatic 270
  steady state 74
  total 78
  transient 74
Rod in
  compression/tension 159
  flexure 154
Rotor
  flexible 250
  unbalanced 248

s-axis shifting 114
s differentiating 114
Saccule 269
Shock 107, 244, 245
Sonometer 174
Spectral density function (see function)
Spectrum
  continuous 109, 117
  line 100, 108, 234
Speed of propagation 150, 158, 161, 164, 169, 173, 191, 279
Spring constant 8, 23, 38, 51, 72, 127, 145, 214, 218, 222, 228, 252
  non-linear 199
  softening 200
  stiffening 200
Spring with mass 145, 173
Standard deviation 236
Step function (see function)
Stiffness 10
Strain 9, 160, 182
  gauge 93, 231
Stress 9, 17, 180
  shear 48
String 149
Sub-harmonic generation 207
Superposition 117
Symmetry
  axis of (see axis)
  circular 167

Taylor series expansion 150, 157, 159, 164, 190
Temperature 16, 169, 183
Threshold of perception of vibration 267
Time constant or relaxation time (see relaxation time
  differentiating 114
  scaling 113
  shifting 114
Torque 5, 17, 147, 156
Torsion 12
Transfer function, system 119
Transducer 180, 224, 244
Transform
  Fourier (see Fourier transform)
  Laplace (see Laplace transform)
  parameter, s 112
Transmissibility 254, 263, 268, 271, 276, 281
Transmission line 171, 176
Trolley, Fletcher's 18

Underdamping 52, 76 (also see damping)
Utricle 269

Variance 236
Vector
  addition 3, 18
  decomposition 4
Velocity
  angular 5, 17
  linear 2, 17, 38, 49, 174, 281
Vestibular apparatus 267
Vibrometer 93, 225, 229, 263, 275, 281
Viscosity 47
  coefficient of (see coefficient)
  of blood 277
Voltage 37, 64, 92, 139, 171, 195, 224, 231, 275, 282 (see also potential)
Volume 169

Wave
  dispersive 158
  equation 150, 157, 161, 164, 168, 191
  forward travelling 151, 191, 280
  reverse 151, 191
  standing 153, 162
Wavelength 153, 175, 195
Wheatstone bridge 232
Work 13, 17

Young's modulus (see modulus)